CONVECTION OVEN COOKERY

Revised Edition

Christie Katona
Thomas Katona

BRISTOL PUBLISHING ENTERPRISES
Hayward, California

A nitty gritty® Cookbook

Printed in the United States of America.

ISBN: 1-55867-264-8

Cover design: Frank J. Paredes
Cover photography: John A. Benson
Food stylist: Susan Devaty
Illustrations: Jim Balkovek

CONTENTS

CONVECTION OVENS:
HOW THEY WORK; HOW TO USE THEM

Convection ovens cook food more rapidly, brown more evenly, and use about 50% less energy than conventional ovens. More importantly, the food is more palatable. Roasted meats come out beautifully browned and crispy on the outside while remaining moist and succulent on the inside. Baked goods brown more evenly and leavened bakery goods rise higher and develop a beautiful golden brown crust. No added fats are necessary when cooking in a convection oven, and excess fat in the food drains away as food cooks.

Commercial bakeries and restaurants have been using convection ovens for more than 25 years. Convection ovens have also been available to the general public for years, but only recently have become well accepted.

The success of convection ovens in the last few years is partly due to the fact that convection oven manufacturers are aggressively promoting their products and partly to the fact that the public is very interested in any appliance that saves them time in the kitchen and that provides a healthy way to cook for their families.

The convection oven has saved us time and fuss in the kitchen and provided many delicious meals. We hope you enjoy the recipes as much as we have. Bon appetit!

HOW CONVECTION COOKING WORKS

Household convection ovens come in various shapes and sizes, but the basic principle is the same for all. A high-speed fan forces air past an electric heating element. The heated air is continuously circulated around all sides of the food which is placed on elevated racks. This provides even, efficient cooking, without hot spots.

Meats and poultry are quickly seared on all sides at the same time, which seals juices in and provides beautiful browning. Yeast breads, cakes, and other leavened bakery items tend to rise higher because the swirling action of the air reduces the external pressure, thus allowing the gas produced by the leavening agent to expand further than it would in a conventional oven.

OTHER CONVECTION OVEN BENEFITS

- Provides the browning capabilities of conventional ovens, but with the shorter times associated with microwave ovens.
- Performs the function of many different kitchen appliances; range, toaster oven, broiler, fryer, steamer, grill and rotisserie.
- Reduces cooking time from 20% to 50%, depending on the food.
- Uses about half the energy of conventional ovens due to the smaller volume and because forced air circulation reduces cooking time.

- Costs less than conventional ovens.
- Eliminates the need to baste and turn most food items.
- Cooks an entire meal at once in the same oven.
- Allows you to see your food cooking in most models.
- Promotes healthy eating: you can air fry to provide the crisp taste of fried foods with much less fat.
- Countertop models take limited space, are compact, lightweight.
- Countertop models provides portability: can be used at home, office, dormitory, or while traveling.
- Countertop models allow easy disassembly and cleaning.

COMPARISON OF CONVECTION OVENS

In the preparation of this book we used six different convection ovens from five different manufacturers. All provided excellent cooking results; however, the round countertop models were set apart by capacity, convenience features, materials, and of course, price. The manuals that came with the units were all well written with clear illustrations and easy instructions on setting up the oven and the use of accessories. Be sure to read carefully the manufacturer's instructions that came with your convection oven.

COOKING VARIABLES

The recommended times and temperatures we have included are good guides. Nonetheless, experienced cooks know to rely on their own instincts and experience. Sometimes the food's appearance or the use of an instant-read thermometer are the best methods. There are many variables which can affect the cooking time and we have listed a few below:

- food temperature prior to cooking
- the amount and shape of the food — a thin or narrow cut of meat will cook much faster than a thick piece of the same weight
- variations in the fat or moisture content of the food
- voltage variations in your utility company's power
- the container used: dark containers absorb heat while shiny metal containers reflect heat

HELPFUL HINTS AND TECHNIQUES

GENERAL TECHNIQUES

Distribute food evenly to ensure an even airflow around the food, typically 1 inch between foods and oven surfaces.

When cooking individual items, arrange them around the outside of the rack to provide unimpeded airflow from the fan through the center of the oven.

Cut large meats into smaller pieces to save time and energy.

When cooking small or lightweight foods, use the hold-down racks supplied by the manufacturer or use a lower fan setting to prevent foods from being blown around by the fan.

To check for doneness, an instant-read thermometer is probably the safest for meats and poultry. Place a meat thermometer in the thickest part of the meat, being careful not to touch a bone or insert it into a fat pocket. The usual methods for conventional ovens also apply to convection cooking: appearance; the cake surface springs back when pressed slightly; a toothpick comes out clean from baked goods; the drumstick moves easily on poultry.

ROASTING/BROILING

Select meats with some marbling, or marinate less expensive cuts of meat prior to cooking.

Place the uncovered roast or poultry on a rack to ensure even circulation of hot air under the meat.

Brush meats, poultry, or fish with melted butter, margarine, or oil to help retain moisture and to aid in browning.

Remember that when starting with frozen meat, cooking times will take approximately 50% longer than fresh meat.

For roasts, place the fat side up so that the melted fat will baste the meat and keep it moist, and for thick cuts, turn at the half-way point to provide even basting. Similarly, when roasting large poultry, place the breast side down for the first half of the cooking cycle to allow the meat juices to baste the breast meat. At the half-way point, turn the breast side up. You may also want to cover the drumsticks with foil at this point to prevent excessive browning.

If you will be carving your roast or turkey, it is best to let it stand for about 15 minutes after cooking. This will make it easier to carve and will also help to retain juices.

BAKING

Remember to leave at least 1 inch between pans and the oven surfaces to provide the needed airflow for even baking.

Metal baking pans tend to provide better results than glass or ceramic dishes. Shiny pans reflect heat and tend to produce lighter and more tender crusts. Dark pans absorb heat and tend to produce browner and more crisp crusts. If you choose to use glass or ceramic dishes, you may want to place them on an aluminum cookie sheet or use a shaped piece of aluminum foil to provide better heat distribution.

Cakes cook faster in tube or Bundt pans due to the greater amount of exposed pan surface.

DEFROSTING

Your convection oven allows you to defrost frozen foods more evenly than a microwave oven. Use the lowest setting (125° on the models we used) and check every 10 minutes, or sooner if you are defrosting a small frozen item.

STEAMING

If your oven comes equipped with a perforated steamer tray, put ½ cup of water in the bottom of the oven and place vegetables or other foods in the steamer tray. You can also add a collapsible steamer basket or any device, such as a tuna fish can with both ends removed, to hold foods out of the water. To steam foods while you are baking or roasting other items, seal the food in foil with a small amount of water.

CLEANING

Although the ovens we used were easily cleaned, a little preparation before cooking can save cleanup time and fuss. Prior to cooking, season the racks with light cooking oil and heat on high for several minutes. When you are ready to cook, spray a little nonstick coating like PAM on the racks and dishes. Alternatively, buy and use

Silverstone precoated racks and dishes (our preference). If you use these nonstick accessories, be careful not to use oven cleaners which can damage the coating.

In general, use hot sudsy water and a plastic scouring pad to clean the oven surfaces and racks. Do not use abrasive pads or abrasive cleaners since they will scratch plastic surfaces and damage the nonstick surfaces.

If you have stubborn, caked-on food on the bottom of the oven, you can use the oven to help you clean. Place about $1/2$ to $3/4$ inch of hot sudsy water on the bottom of the oven, set the oven temperature to 200°, and set the timer to 10 or 15 minutes.

Convection ovens have air filters which must be periodically cleaned. The frequency of cleaning depends on the model you have and the type of cooking you do. As a general recommendation, clean the filters when you can no longer easily see light through the filters when held up to the light. Soaking in hot sudsy water and rinsing will usually be sufficient. For stubborn grease, some manufacturers recommend soaking in a commercial product like Dip-it. Follow the manufacturer's recommendations that come with your oven.

ADAPTING RECIPES

Just about any recipe that you have for conventional ovens will also work in your convection oven, although you will generally have to reduce the time or temperature or both. The manufacturer of the oven usually supplies a fairly complete table of foods with recommended times and temperatures which you can compare to your own recipes. Alternatively, you can use the following general rules:

For everything except recipes that contain leavening agents like yeast, baking soda, eggs, etc., you should decrease the recommended temperature by 50° and the cooking time by ⅓.

For leavened items like breads, rolls, cakes, etc., use the same time and decrease the temperature by 75°, but not below 300°.

APPETIZERS

QUICK CRAB DIP

If you have a can of crabmeat in your cupboard, you are always ready for spur-of-the-moment guests.

8 oz. cream cheese
1 tbs. milk
2 tsp. Worcestershire sauce
1 cup crabmeat
2 tbs. chopped green onions
2 tbs. sliced almonds

Heat oven to 300°. With a mixer or food processor, combine cream cheese, milk and Worcestershire sauce. Gently stir in crabmeat and green onion. Top with almonds. Place in a shallow buttered ovenproof glass baker, such as an au gratin dish or quiche plate. Bake for 10 to 15 minutes or until hot and almonds are toasted. Serve hot with crackers for dipping.

SAVORY ONION CHEESECAKE

Servings: 16-20

You can make this delicious appetizer up to three days before your party.

CRUST

1 1/4 cups crushed Ritz cracker crumbs
1/4 cup butter, melted

Heat oven to 350°. Combine crust ingredients and press into the bottom of a 9-inch springform pan. Bake for 5 to 8 minutes or until browned. Set aside while making filling.

FILLING

1 tbs. butter
3 large yellow onions, peeled and
   thinly sliced
1 tsp. sugar
12 oz. cream cheese, cubed
3 eggs

2/3 cup sour cream
generous dash Tabasco Sauce
4 oz. Swiss cheese, shredded
4 oz. Monterey Jack cheese, shredded
2 tbs. freshly grated Parmesan cheese
paprika and fresh parsley for garnish

In a heavy skillet, melt butter and sauté, onions until soft and golden. Sprinkle with sugar. Combine cream cheese, eggs, sour cream and Tabasco. Place onion mixture in prepared crust. Sprinkle with shredded cheeses. Pour cream cheese mixture over onion mixture. Sprinkle with Parmesan cheese.

Reduce oven setting to 325° and bake for 15 to 20 minutes or until set, testing doneness with a toothpick. Cover and refrigerate. To serve, remove sides of pan and place on a pretty serving dish, sprinkle with paprika and garnish with parsley. Cut into thin wedges.

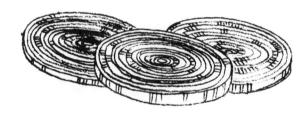

SMOKED SALMON CHEESECAKE

Savory appetizer cheesecakes make an easy do-ahead dish for a crowd. For variation, use crumbled bacon and finely chopped spinach or blue cheese with bacon and sautéed onion.

CRUST
2 tbs. butter
1/3 cup fine dry French breadcrumbs
1/4 cup grated Swiss cheese
1/2 tsp. dried dill

Butter a 9-inch springform pan. Mix breadcrumbs, cheese and dill. Sprinkle crumb mixture into pan, turning to coat bottom and sides. Chill while making filling.

FILLING

3 tbs. butter
1 medium onion, finely chopped
1 1/3 lb. cream cheese, room temperature
4 eggs
1/2 cup grated Swiss cheese
1/3 cup half-and-half
1/2 tsp. salt
8 oz. smoked salmon, coarsely chopped

Heat oven to 300°. Melt butter in a heavy skillet and sauté, onion until soft. With a mixer or food processor, combine cream cheese, eggs, Swiss cheese, half-and-half and salt until smooth. Fold in onion and salmon. Filling should have texture. Pour into prepared pan. Set pan in another larger pan and fill halfway up sides with hot water to form a water bath. Bake for 1 hour and 10 minutes or until firm near center.

Turn off oven and let cheesecake cool in oven to room temperature. Cover and refrigerate. To serve, bring to room temperature and slice into wedges.

APRICOT WINGS

Chicken wings make a sure hit as an appetizer. This sauce is also good on spareribs or as a dipping sauce for figs or dates wrapped in bacon and skewered with a toothpick.

3 lb. chicken wings
salt and pepper
2 jars (4 oz. each) apricot or peach baby food
1/3 cup catsup
1/3 cup cider vinegar
2 tbs. soy sauce
1/2 cup brown sugar, packed
1 clove garlic, minced
1 tbs. grated fresh ginger

Place chicken wings in an ovenproof casserole. Combine remaining ingredients and pour over wings. Heat oven to 325° and bake wings for 45 to 60 minutes, basting with sauce occasionally. Can be prepared ahead and reheated.

BLACK-EYED SUSANS

One of our favorite appetizers ever! These are particularly nice for a brunch. If you can find them, Medjool dates are very nice.

1 cup butter
1 lb. sharp cheddar cheese, shredded
2 cups flour
1 tsp. salt
1 tsp. cayenne pepper
1 lb. dates, pitted and halved lengthwise
confectioners' or granulated sugar for dusting, optional

Heat oven to 300°. With a mixer or food processor, combine butter, cheese, flour and seasonings to form a soft dough. Roll out on a lightly floured surface and cut into 3-inch rounds. Place date on one side and fold over to make a crescent shape. Press edges with the tines of a fork to seal. Place on a pizza pan. Bake for 20 to 30 minutes or until golden. Remove from baking sheet and cool on wire racks. Dust with confectioners' sugar or granulated sugar, if desired. Serve warm or at room temperature. These can be made ahead and reheated, and also freeze well.

ARTICHOKE AND CRAB BAKE

Expensive but worth it—elegant served in a chafing dish.

8 oz. cream cheese, room temperature
1 cup mayonnaise
1 small onion, diced
2 tbs. butter
1 can (10 oz.) artichoke hearts, drained and chopped
1 cup fresh crabmeat or baby shrimp
dash Tabasco Sauce
1/3 cup freshly grated Parmesan cheese

Heat oven to 350°. With a mixer or food processor, combine cream cheese and mayonnaise. Sauté, onion in butter until soft. Add to cream cheese mixture with artichokes. Gently stir in crab by hand. Season to taste with Tabasco. Spoon into an ovenproof serving dish and top with Parmesan cheese. Bake for 15 to 20 minutes or until bubbly. Serve with sliced French bread or crackers.

QUICK QUICHE SQUARES

These are quick because they don't have a crust. They can be frozen after baking, but be sure to separate layers with waxed paper before freezing.

1 lb. kielbasa sausage, chopped
1 medium onion, chopped
6 eggs
16 oz. cream cheese, room temperature
1 1/2 cups sour cream
1/2 cup flour
3 cups shredded sharp cheddar cheese
1/3 tsp. white pepper

Heat oven to 325°. Chop kielbasa and onion. Sauté in a skillet until onion is tender. Drain off any fat. Combine remaining ingredients with a mixer or food processor. Add kielbasa and onion and mix well. Butter two 9-inch square baking pans. Pour mixture into pans, dividing evenly. Bake for 20 to 25 minutes or until set and tops are golden. Cool and cut into 1 1/2-inch squares.

MYSTERY MEATBALLS

Servings: 8-12 as appetizer

The mystery of these meatballs is the unusual ingredients in the sauce. They make a great appetizer served in a chafing dish at holiday time.

1 clove garlic, minced
1 tbs. finely chopped fresh parsley
2 green onions, finely chopped
1/2 cup breadcrumbs
1 can (8 oz.) water chestnuts, drained and coarsely chopped
2 eggs, beaten
1 tsp. salt
1/2 tsp. pepper
1/3 lb. pork sausage
1 lb. ground beef

Heat oven to 325°. Combine garlic, parsley, green onions, breadcrumbs and water chestnuts. Add eggs, seasoning and meats and blend thoroughly. Form into 1-inch balls and place in a shallow baking pan (a pizza pan works well). Bake for 12 to 18 minutes or until cooked through. Drain off any juices. Make sauce, add meatballs to sauce and heat through.

SAUCE

1 can (16 oz.) jellied cranberry sauce
$\frac{1}{2}$ cup brown sugar, packed
$\frac{1}{4}$ cup prepared barbecue sauce
1 tbs. Dijon mustard

In a large saucepan, heat ingredients over medium heat, stirring until smooth.

MUSHROOM PIZZA

This makes a nice appetizer or a casual supper when served with a green salad.

1 tube (8 oz.) crescent rolls
6 tbs. butter
1 clove garlic, minced
1 onion, chopped
1 lb. mushrooms, sliced
1 tsp. lemon juice
1/2 tsp. dried basil leaves

1/2 tsp. dried oregano leaves
1/2 tsp. black pepper
1/4 cup red wine
1 pkg. (8 oz.) cream cheese, room
   temperature
2/3 cup shredded Parmesan cheese

Heat oven to 300°. Spray a pizza pan with nonstick cooking spray. Unroll crescent rolls and separate into triangles. Press them out on the pizza pan with points facing the center. Press any perforations together.

Melt butter in a large skillet over medium heat. Add garlic and onion and sauté, until onion is tender. Add mushrooms and seasonings and cook until mushrooms are tender. Add wine, turn heat to high and cook until wine is evaporated.

Gently spread cream cheese over dough. Top with cooked mushroom mixture and sprinkle with Parmesan cheese. Bake for 20 to 25 minutes or until crust is golden and topping is bubbly. Cut into wedges to serve.

GOAT CHEESE-STUFFED MUSHROOMS

Makes 3 dozen

This upscale hors d'oeuvre is sure to impress. Select mushrooms just a bit over 1 inch in diameter.

36 white mushrooms, stems removed
1/4 cup olive oil
1 cup chopped onion
1/2 lb. bacon, cooked crisp and crumbled
1 pkg. (10 oz.) frozen chopped spinach, thawed and squeezed dry
8 oz. goat cheese (chevre)
2 tbs. cream
1 tsp. salt
1/2 tsp. ground nutmeg
1/2 tsp. pepper

Heat oven to 325°. In a large skillet over medium-high heat, cook mushroom caps until golden. Remove and set aside. Cook onion in same skillet until softened. In a bowl, combine onion, spinach, bacon and goat cheese. Stir in cream and seasonings. Divide stuffing mixture evenly into mushroom caps. Place on a baking sheet and bake for 5 to 8 minutes. Serve warm. Can be made ahead and refrigerated until ready to back.

STUFFED MUSHROOMS

There are dozens of delicious combinations a person can create given two dozen mushrooms and a bit of cream cheese. Listed below is our standard technique and some of our favorite combinations. The mushrooms soak up that lovely garlic butter, and it adds a lot of flavor.

2 tbs. butter
1 clove garlic, minced
24 large mushroom caps
4-8 oz. cream cheese with choice of filling
   variations, follow

Heat oven to 325°. In a shallow ovenproof casserole, melt butter with garlic. Add mushroom caps round-side down and bake for 5 minutes while blending cream cheese with a selected filling variation. Remove casserole from oven and spoon filling into caps. Continue baking for 5 to 10 minutes longer or until piping hot. Serve on small plates or with plenty of napkins.

VARIATIONS

- cream cheese, chopped frozen spinach (squeeze very dry), sautéed onion and crumbled bacon, topped with cheddar cheese
- cream cheese and commercial liver paté (it's good! even if you don't like paté)
- cream cheese and $1/2$ cup prepared pesto, topped with Parmesan cheese
- cream cheese, chopped ham, green peppercorns and walnuts
- cream cheese, shrimp, green onions and dill, topped with Swiss cheese
- cream cheese, chopped artichokes and crab, topped with Jack cheese
- cream cheese, green chiles, chopped black olives and green onions, topped with cheddar cheese
- cream cheese, sun-dried tomatoes, chopped salami and green pepper, topped with fontina cheese

BRIE EN CROUTE

For a stunning appetizer, it's hard to beat Brie en Croute. We like to serve it at special parties on our prettiest silver tray and have a beautiful red rose peeking out under the handle. This recipe is also delicious using orange marmalade instead of jam. The combination of warm rich cheese, tangy jam, salty prosciutto and pastry is unbeatable.

1 large Brie, about 2 lb.
1 1/2 cups raspberry jam
1 pkg. Pepperidge Farm puff pastry sheets, thawed
4 oz. prosciutto ham, sliced paper thin
1 egg, beaten
1/4 tsp. salt

Leaving rind on Brie, cut chilled cheese in half horizontally. Spread cut surface with raspberry jam. Replace top. Roll puff pastry out on a lightly floured board. Place Brie in center. Cut ham into strips 1 inch wide and place on top of Brie. Roll out other piece of puff pastry and place on top of cheese. Press top sheet down to meet bottom sheet and trim in a circle leaving a 1-inch border around cheese. Press edges together firmly using a fork. Place cheese on a pizza pan which has been sprayed with nonstick cooking spray. Fold edges of puff pastry underneath cheese, being careful not to stretch or tear pastry.

Using trimmed scraps from pastry, cut decorative flowers, leaves or other patterns to decorate top of cheese. Beat egg with a pinch of salt. Brush cheese with egg glaze. Use egg glaze to adhere decorations to cheese and brush these as well. Can be refrigerated at this point. Cover well with plastic wrap.

To bake, heat oven to 350° and bake cheese for 25 to 30 minutes or until puffed and golden. Let stand at room temperature for at least 1 hour before serving. It is best not to transfer to another serving tray: put pizza pan in the center of a prettier platter and arrange crackers around the edge of the cheese for serving. To serve, cut into wedges and serve with water crackers.

BAKED BRIE WITH APPLES, PECANS AND KAHLUA

Here is one of our favorite holiday appetizers, served with water crackers. You can adapt this to any size Brie, large or small.

1 Brie cheese, about 16 oz.
2 Granny Smith apples, quartered, peeled and chopped
$1/2$ cup pecans
$1/3$ cup brown sugar, packed
2 tbs. Kahlua liqueur

Place Brie in a shallow ovenproof dish or a pizza pan. Mix chopped apples with nuts, sugar and Kahlua. Apple pieces should be about $1/4$ inch. Spread on top of cheese. Heat oven to 300°. Bake cheese for 10 to 20 minutes or until topping is bubbly and cheese is softened. Time will depend on temperature and ripeness of cheese as well as size.

ROSEMARY WALNUTS

These make a lovely gift from your kitchen. They freeze well, make a nice nibble with cocktails or can be added to a tossed salad. One of our favorites is greens, shredded Gruyère cheese, Rosemary Walnuts, avocado and red onion slices with a vinaigrette—different and good.

6 tbs. butter
1 tbs. dried rosemary
1 tbs. salt
$1/2$-$1/3$ tsp. cayenne pepper
4 cups walnut halves

Heat oven to 325°. Melt butter in a 9-inch pan. Stir in rosemary, salt and cayenne. Add walnuts and stir to coat well. Bake for 10 to 15 minutes or until brown. They begin to smell wonderful as they approach being done.

APPETIZER CUPS

Miniature muffin pans have to be one of a cook's best friends! They make appetizers look stunning and are just the right size for parties. Be sure to get the ones with a nonstick coating. Here are a couple of ideas for crisp and golden cups to fill with any of the mushroom filling suggestions on page 25, or have fun with your own creations. Won ton wrappers are usually available in the produce section of your grocery store.

WON TON CUPS

1 pkg. won ton wrappers, square or round
$1/2$ cup butter

Heat oven to 325°. Spray several muffin pans with nonstick spray. Brush wrappers lightly with butter and press into individual cups. Bake for 5 minutes or until pale gold. Fill with desired filling and sprinkle tops with cheese. Continue baking for 5 minutes longer or until filling is heated through and cheese is melted.

PHYLLO CUPS

These are more fragile than Won Ton Cups, but have that melt-in-your-mouth quality. They make a nice dessert when filled with cream cheese combined with grated orange zest and confectioners' sugar—top with a strawberry, raspberry or blueberries. At holiday time use a bit of whole berry cranberry sauce. Festive! Fill the cups right before serving.

8 sheets phyllo
1/4 cup butter, melted

Brush phyllo sheets lightly with butter and stack 4 sheets high. Using scissors, cut phyllo into 4 strips lengthwise and 6 strips crosswise. You should have twenty-four 2-inch squares. Press each into a miniature muffin cup. Heat oven to 325°. Bake for 5 minutes. Fill as desired and sprinkle with cheese. Continue baking for 3 to 5 minutes longer until cheese is melted. For dessert cups, bake until golden, remove from pans and cool. Fill with cream cheese just before serving. They become soggy if filled and left standing. Do not use a pastry bag, as a spoon is much easier.

YEAST BREADS AND QUICK BREADS

LOAFER'S LOAF

This recipe is really easy, particularly if you have a food processor. Experiment with ingredients. Add various seeds—sesame, poppy or sunflower. Substitute cracked wheat, whole wheat or oats for about 1/3 of the total flour required.

1 tbs. dry yeast	3 tbs. butter
1 cup flour	1 tsp. salt
1 1/4 cups hot water (120°)	2 cups flour
1 tbs. sugar	cornmeal

With a food processor, combine yeast and 1 cup flour. Mix water and sugar together and stir until dissolved. With motor running, pour liquid down feed tube. Let machine run for 1 minute. Add butter, salt and remaining flour and let machine run until dough forms a ball. Put dough in a greased bowl. Cover and let rise in a warm place for 1 hour. Punch down, cover and let rest for 10 minutes.

Grease a round casserole dish or ovenproof bowl and sprinkle it generously with cornmeal. Put dough in bowl, cut several slashes in the top, and sprinkle with additional cornmeal. Let rise for 45 minutes.

Heat oven to 350°, and bake bread for 30 to 40 minutes or until loaf sounds hollow when tapped. Cut into wedges to serve or slice across the loaf.

ONION ROLLS

The smell of these rolls baking is heavenly. Half of the onions go into the rolls and the other half go on top with a glaze.

2 large onions, chopped
1/4 cup butter
1 pkg. active dry yeast
1/4 cup warm water
1 tsp. sugar
1/2 cup butter, melted and cooled
1 cup milk
2 cups flour
1/4 cup sugar
2–2 1/2 cups additional flour
1 egg
2 tsp. salt

Sauté onions in butter in a large skillet over medium high heat until they are tender and golden brown. Remove and set aside.

With a mixer or food processor, combine yeast, warm water and sugar. Let stand for 5 minutes. Add butter, milk and flour and mix for 2 minutes. Add sugar, flour and half of sautéed onions and beat until mixture forms a soft dough. (It may be necessary to divide the dough in half if you are using a food processor and it starts to slow down.) Place dough in a greased bowl, turning to coat surface. Cover and let rise in a warm place for 1 hour until doubled. Punch dough down.

Grease two 8-inch cake pans. Shape dough into eighteen 2-inch balls and place 9 in each pan. Combine remaining sautéed onions with 1 egg beaten with a pinch of salt. Brush rolls with egg-onion mixture. Let rolls rise again, uncovered, for 45 minutes.

To bake, heat oven to 325°. Bake for 20 to 30 minutes until golden.

CHEESE AND ONION LOAF

Makes 1 loaf

I have always had good results making yeast breads in the food processor. Of course, you can also make this in a mixer with a dough hook or by hand. This bread is a pretty orange color and is delightful for meat loaf sandwiches or with soups.

1 pkg. yeast
1/4 cup warm water
2 tbs. sugar
6 oz. sharp cheddar cheese, shredded
1 carrot, finely shredded

3 tbs. vegetable oil
1 1/2 cups milk
1 tsp. salt
5-6 cups flour

With a food processor or heavy duty mixer, combine yeast, warm water and sugar. Set aside for 5 minutes. Add half of the milk and flour to workbowl and let machine run for 2 minutes to develop gluten in flour. Add cheese, carrot, oil, remaining milk and salt and process until batter is smooth. Add remaining flour until batter forms a smooth dough. Remove from workbowl and knead on a lightly floured board. Place in an oiled bowl, cover and let rise in a warm place until doubled.

Heat oven to 325°. Punch dough down. Shape into a rustic dome and slash top in a large cross. Place on a greased pizza pan. Proceed with topping.

TOPPING

2 tbs. butter, melted
1/2 cup chopped onion
2 tsp. paprika
1 tsp. salt

Brush loaf with melted butter and sprinkle with onion. Press into top. Sprinkle with salt and paprika. Bake for 30 to 45 minutes or until loaf sounds hollow when tapped.

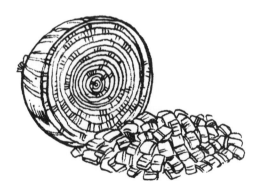

SAVORY TOMATO YEAST BREAD

Makes: 1 loaf

Try a sandwich with cream cheese and sliced turkey or chicken on this unusual bread.

1 pkg. yeast
1/3 cup warm water
2 tsp. sugar
1 cup water
1 cup flour
1 small onion, finely chopped
1/4 cup butter, softened
1 tsp. salt
2 tsp. celery salt
1/3 cup tomato sauce
3-4 cups additional flour

With a heavy duty mixer or food processor, combine yeast, warm water and sugar. Set aside for 5 minutes. Add 1 cup additional water and 1 cup flour and let machine run for several minutes to develop gluten. Add onion, butter, salt, celery salt and tomato sauce. Process to make a smooth batter. Add remaining flour until mixture forms a soft dough. Remove from workbowl and knead for several minutes on a lightly floured board. Place in a greased bowl, turn to coat, and cover. Let rise in a warm place until dough is doubled in bulk.

Punch dough down and shape into a loaf. Place in a greased 9-inch bread pan. Let rise again until dough is 1 inch above top of pan.

Heat oven to 350°. Place pan on center rack. Bake for 20 minutes, remove loaf from pan and continue baking for 5 to 15 minutes longer until loaf sounds hollow when tapped.

POPPY SEED ONION LOAF

This bread, one of our all-time favorites, is great with soup for supper.

DOUGH

1 pkg. active dry yeast
1/4 cup warm water
1 tsp. sugar
1/2 cup butter, melted and cooled
1 cup milk

4–4 1/2 cups flour
1/4 cup sugar
2 tsp. salt
1 egg

With a heavy duty mixer or food processor, combine yeast, water and 1 tsp. of sugar. Set aside for 5 minutes. Add cooled butter, milk and 2 cups of flour and let machine run for 3 minutes. Add remaining ingredients until you have a smooth dough which is fairly stiff. Shape dough into a ball, place in a greased bowl and turn to coat all surfaces. Cover and let rise until doubled, about 1 hour. Make filling and glaze.

FILLING

1 cup chopped onion
$1/4$ cup butter
3 tbs. poppy seeds
$1/2$ tsp. salt
1 egg for glaze

Combine filling ingredients. Set 2 tbs. of the filling aside in a small bowl, add 1 egg and whisk to make glaze.

To assemble loaf: Punch dough down and roll out on a lightly floured surface into a 10 x 15-inch rectangle. Spread with filling and roll up lengthwise as for jelly roll. Press ends closed and fold under about 1 to 2 inches of unfilled dough. Shape dough into a crescent and place on a greased baking sheet such as a pizza pan. Let rise until double, about 1 hour.

Heat oven to 300°. Make several slashes in the top of the loaf. Brush loaf with glaze. Bake for 30 to 40 minutes or until bread sounds hollow when tapped. Cool slightly before slicing.

HARVEST LOAF

This beautiful loaf of bread makes a welcome and dramatic gift. Decorate the top with a sculptured grape cluster—just roll bits of dough into rounds the size of marbles. Roll another bit of dough into a long vine and twist into a spiral. Next cut a few leaves, and press onto the top of the loaf. It's easy!

2 cups warm water
2 pkg. yeast
1/4 cup brown sugar, packed
2 cups white flour

2 tsp. salt
1 cup oatmeal
1/2 cup molasses
3-4 cups whole wheat flour

Using a heavy duty mixer or food processor, combine water, yeast and brown sugar. Set aside for 5 minutes until foamy. Add 2 cups white flour and let machine run for several minutes. Add salt, oatmeal and molasses; process for 1 minute. Add whole wheat flour 1 cup at a time until a stiff dough forms.

Place in an oiled bowl and turn to coat surface. Cover and let rise for 2 hours. Punch down and remove about 1/5 of the dough to use to decorate the loaf. Shape remaining dough into a large round loaf. Place on a greased pizza pan. Arrange grape cluster on top. Let rise until double.

Heat oven to 325°. Place loaf on bottom rack and bake for 45 minutes to 1 hour or until loaf sounds hollow when tapped. Cool.

QUICK DINNER ROLLS

Just about an hour for hot dinner rolls from scratch! Treat your family sometime soon.

1 cup warm water
1 pkg. active dry yeast
2 tbs. sugar
2¼ cups flour
1 tsp. salt
1 egg
2 tbs. butter, melted

With a heavy duty mixer or food processor, combine water, yeast and sugar. Set aside for 5 minutes. Add 1 cup of flour and mix for 2 minutes. Add remaining ingredients and beat to make a smooth dough. Remove dough to a greased bowl, cover and let rise for 30 minutes. Stir dough down and spoon into greased muffin cups. Let rise again until dough reaches tops of muffin cups, about 20 minutes.

Heat oven to 350° and bake for 12 to 18 minutes or until golden.

HERBED GARLIC LOAF

Servings: 6

So good, one of us could eat the whole thing!

1 large loaf sourdough French bread, unsliced
3/4 cup butter
1/4 cup finely chopped fresh parsley
1/4 cup finely chopped green onions
1/2 tsp. salt
1/2 tsp. dried basil
1/2 tsp. dried thyme
2 cloves garlic, minced
1/4 cup grated Parmesan cheese

Heat oven to 375°. Trim crusts from top and sides of bread. Cut into 2-inch diagonal slices in each direction, slicing almost to bottom crust. Combine butter, parsley, onions and seasonings. Spread mixture between cuts and over top and sides of loaf. Sprinkle top with Parmesan cheese. Bake for 15 minutes or until crusty brown.

GREAT GARLIC TOAST

Feel free to use any combination of cheeses you might have on hand.

1 long loaf French bread, unsliced
1 cup butter, softened
3 cloves garlic, minced
1 cup shredded cheddar cheese
1 cup shredded Monterey Jack cheese
1 cup shredded Parmesan cheese

Heat oven to 400°. Slice bread in half horizontally. In a small bowl, combine butter and garlic. Spread on cut sides of bread. Sprinkle evenly with cheeses. Place bread on center rack, cutting in halves if necessary. Bake until cheeses melt and bread is piping hot. Slice on the diagonal to serve.

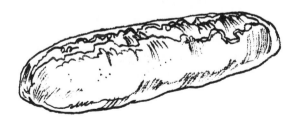

CARAMEL PECAN ROLLS

Dainty and easy to fix, these are just the thing to serve for a special breakfast.

5 tbs. butter, melted
3/4 cup brown sugar, packed
1/4 cup water
1/2 cup chopped pecans
2 cans (8 oz. each) crescent rolls
3 tbs. butter
1/4 cup sugar
2 tsp. cinnamon

Heat oven to 325°. Combine butter, brown sugar and water. Pour into a 9-inch baking pan. Sprinkle with pecans. Separate each can of crescent rolls and divide each into 4 rectangles; seal perforations. Spread with 3 tablespoons butter, and sprinkle with sugar and cinnamon. Roll up each rectangle from the short side. Cut each roll into 3 pieces and place in prepared pan. Bake for 15 to 20 minutes or until rolls are puffed and brown. Invert immediately onto a platter. Spoon any sauce remaining in pan over top of rolls. Serve hot.

GOOEY APPLESAUCE COFFEECAKE

Servings: 12

The topping and applesauce in this coffeecake meet to make moist rich tunnels.

1/2 cup butter
2 eggs, beaten
2 cups applesauce
2 cups sugar
1 cup buttermilk

3 1/2 cups flour
2 tsp. baking soda
2 tsp. cinnamon
1 1/2 tsp. salt
Topping, follows

Heat oven to 300°. Using a heavy duty mixer or food processor, combine butter, eggs, applesauce, sugar and buttermilk. Add flour and spices. Butter a large baking pan or two 9-inch cake pans. Make topping. Pour batter into prepared pan(s) and sprinkle evenly with topping. Bake for 30 to 40 minutes for 1 large pan and 20 to 25 minutes for 2 pans. Serve warm.

TOPPING

1 cup brown sugar, packed
1 cup granulated sugar
1/2 cup flour

1/2 cup butter
1 tbs. cinnamon
1 cup chopped walnuts or pecans, optional

Combine sugars and flour. Cut in butter, cinnamon and nuts.

PEAR AND PECAN COFFEECAKE

The yogurt in the batter makes this coffeecake moist and tender.

BATTER

1/2 cup butter or margarine
1 cup sugar
3 eggs
2 cups flour
1 tsp. baking powder
1 tsp. baking soda

1/2 tsp. salt
1 tsp. cinnamon
1 cup plain or vanilla yogurt
2 cups chopped peeled pears
1 tsp. vanilla extract

TOPPING

1 cup brown sugar, packed
2 tbs. flour
1 1/2 tsp. cinnamon

1/2 tsp. nutmeg
1/4 cup butter, melted
1 cup chopped pecans

Heat oven to 350°. Grease an 8 1/2-x-11-inch baking pan or a 10-inch round pan. With an electric mixer, cream butter and sugar. Add eggs, beating well. Sift dry ingredients together and add to butter-egg mixture. Stir in yogurt, pears and vanilla. Pour into prepared pan. Stir topping ingredients together and sprinkle evenly over batter. Bake for 25 minutes or a toothpick inserted in the center comes out clean. Let stand 10 minutes before serving.

BUTTER PECAN MUFFINS

Makes 1 dozen

This recipe uses a different method of making muffins in that the batter is prepared the night before and paper liners should not be used.

1/2 cup butter
1 cup brown sugar, packed
1 egg
1 cup milk
1/2 tsp. baking soda
1 tsp. vanilla extract
2 cups flour
1/2 cup chopped pecans

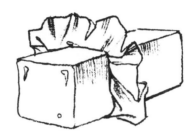

Generously grease 12 muffin cups. Combine butter, sugar, egg and milk until smooth. Add soda, vanilla, flour and nuts. Pour into prepared pan. These can be baked immediately or covered and refrigerated overnight. Heat oven to 350° and bake for 15 to 20 minutes or until muffins are puffed and golden. Serve warm.

FRUITED BRAN MUFFINS

When your bananas get too ripe, they can be frozen in their skins. To use, just thaw briefly, slip them out of the skins and proceed with your recipe.

1 cup raisin bran cereal
1/2 cup bran cereal
3/4 cup orange juice
1 egg, beaten
1/4 cup vegetable oil
1 banana, mashed
1/2 orange, peel and all, grated

1 cup flour
2 1/2 tsp. baking powder
1/2 tsp. salt
2 tsp. cinnamon
1/2 cup sugar
1 cup raisins

Heat oven to 350°. In a large bowl, combine cereals, orange juice, egg, oil and banana. Grate orange, removing any large pieces of membrane, but do use the peel and all. Add orange to mixture. Add flour, baking powder, salt, cinnamon and sugar and stir just until moistened. Fold in raisins. Line muffin cups with paper liners and fill 2/3 full with batter. Bake muffins for 15 to 20 minutes or until done and a toothpick inserted into the center comes out clean.

MORNING GLORY MUFFINS

These muffins are loaded with good things.

2 cups flour
2 tsp. baking soda
2 tsp. cinnamon
1/2 tsp. salt
1v cups sugar
3 eggs
1 cup vegetable oil

1 tsp. vanilla extract
1 cup chopped pecans
1 cup raisins
1 cup coconut
1 cup grated carrots
1 cup grated apple

Heat oven to 300°. Combine flour, baking soda, cinnamon, salt and sugar. Add eggs, oil and vanilla. Add remaining ingredients and mix thoroughly. Line 18 muffin cups with paper liners and fill 2/3 full with batter. Bake for 15 to 20 minutes, or until muffins test done.

LOW-FAT CRANBERRY MUFFINS

Makes 12

These muffins have only 2 grams of fat and 169 calories each!

1 cup skim milk
1 cup oatmeal
1/4 cup unsweetened applesauce
1 tbs. canola oil
2 tsp. grated orange zest
1 tsp. vanilla extract
1/2 cup egg substitute
1 1/2 cups flour

1/2 cup sugar
2 tsp. baking powder
1 tsp. baking soda
1/2 tsp. salt
1 tsp. cinnamon
1 cup dried cranberries
1/2 cup golden raisins

Heat oven to 375°. Spray a 12-cup muffin pan with nonstick cooking spray. Combine milk, oatmeal, applesauce, oil, zest, vanilla and egg substitute in a bowl. Stir together dry ingredients in another bowl. Add dry mixture to wet mixture. Fold in dried cranberries and raisins. Divide evenly into prepared muffin cups. Bake for 15 to 20 minutes or until a wooden pick inserted into the center of a muffin comes out clean. Can be frozen.

BRUNCH

BAKED SWISS FONDUE

Servings: 6-8

Fondue also makes a lovely supper in front of the fire on a cold evening. Add a crisp green salad and a glass of chilled white wine.

$1/2$ cup butter
2 cloves garlic, minced
1 loaf (10 oz.) French bread
6 oz. Swiss cheese, shredded
3 tbs. finely chopped onion
1 tsp. seasoned salt
$1/2$ tsp. dry mustard
1 tsp. paprika
$1/3$ cup butter
$1/4$ cup flour
3 cups milk
1 cup dry white wine
3 eggs, beaten

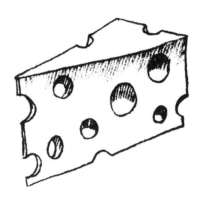

Combine butter and garlic. Remove ends from French bread and cut into $\frac{1}{2}$-inch slices. Spread garlic butter on 1 side of each slice. Butter a 2-quart casserole, and line bottom and sides of casserole with bread slices, buttered side down. Reserve remaining slices.

Toss Swiss cheese with onion, salt, mustard and paprika. In a saucepan, melt remaining $\frac{1}{3}$ cup butter, add flour and stir to form a roux. Gradually add milk and bring to a boil over medium high heat, stirring constantly. Add wine to pan. In a small bowl, beat eggs. Add a small amount of milk and wine mixture to eggs and mix. Pour eggs back into pan, stirring well. Remove from heat. Pour some of the sauce into casserole, sprinkle with cheese mixture and add a layer of buttered bread. Continue layering, ending with bread, buttered side up. Cover and refrigerate overnight.

Heat oven to 325° and bake for 30 to 45 minutes or until puffy and golden.

OVERNIGHT CUSTARD FRENCH TOAST

Servings: 4-8

This recipe comes in handy if you have overnight house guests. For a change of pace, slit a small pocket in the side of each slice of bread and fill with cream cheese, marmalade and nuts, blueberry pie filling or Brie and sliced strawberries.

1 loaf (10 oz.) French bread, cut into 2-inch slices
8 eggs
3 cups milk
1 tbs. vanilla extract
3/4 tsp. salt

Butter a shallow casserole. Trim ends from bread and place slices in dish. Combine remaining ingredients. Pour over bread. Cover and refrigerate overnight. The next morning, heat oven to 325°. Dot top of soaked bread with butter. Bake for 30 to 40 minutes or until puffy and golden. Serve with maple syrup and assorted preserves.

HAM QUICHE WITH POTATO CRUST

Here is something unusual to serve for brunch or a special Sunday breakfast. Add fruit and muffins to complete your menu.

1 pkg. (24 oz.) frozen hash browns, thawed and squeezed dry
1/3 cup butter, melted
1 cup shredded hot pepper cheese
1 cup shredded Swiss cheese
1 cup diced cooked ham
2 eggs, beaten
1/2 cup half-and-half
1/4 cup sliced green onions
1/2 tsp. seasoning salt

Heat oven to 400°. Press potatoes between paper towels to remove moisture. Place in bottom and up sides of a 10-inch metal pie pan to form crust. Brush with melted butter, especially the top edges. Bake for 20 to 30 minutes or until crisp and golden. Remove from oven and sprinkle with cheeses and ham. Beat eggs with cream, onions and seasoning salt. Pour over ham and cheese mixture. Set oven at 325° and continue baking for 25 to 35 minutes or until eggs are set. Cut into wedges like a pie to serve.

LOW-FAT ONION QUICHE

You can make this recipe even lower in fat by eliminating the crust completely.

1 pie crust for 9-inch pie
2 tbs. butter or margarine
2 cups thinly sliced onions
1 cup nonfat plain yogurt
1/4 cup low-fat mayonnaise
3/4 cup egg substitute
1 tbs. chopped fresh chives
1/2 tsp. dried thyme
1 tsp. salt
1/2 tsp. pepper
1/4 cup freshly grated Parmesan cheese

Heat oven to 325°. Bake pie shell for 5 minutes and set aside. Melt butter in a heavy skillet over medium high heat. Cook onions stirring, until they begin to soften. Place in prepared pie shell. Mix together the yogurt, mayonnaise, egg substitute and seasonings. Pour over onions. Bake for 40 minutes or until the center begins to firm. Sprinkle with Parmesan and continue baking for 10 minutes longer. Cut into wedges to serve.

BAKED HAM WITH CITRUS SOUFFLÉ

A light and tangy topping crowns this showy entrée.

1 cooked ham slice, about ¾-inch thick
2 egg whites, room temperature
1 green onion, finely chopped
2 tbs. orange juice concentrate
2 tbs. brown sugar
2 tbs. dried currants
1 tsp. orange zest

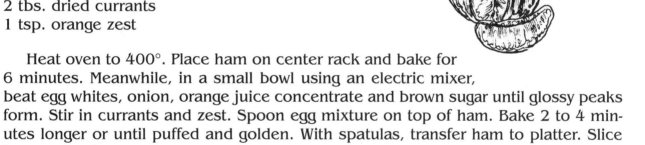

Heat oven to 400°. Place ham on center rack and bake for
6 minutes. Meanwhile, in a small bowl using an electric mixer,
beat egg whites, onion, orange juice concentrate and brown sugar until glossy peaks
form. Stir in currants and zest. Spoon egg mixture on top of ham. Bake 2 to 4 minutes longer or until puffed and golden. With spatulas, transfer ham to platter. Slice
ham crosswise to serve.

BAKED BLINTZ CASSEROLE

Servings: 8

This is a wonderful make-ahead brunch dish for a crowd. You can assemble it the day before your party and refrigerate. Serve an assortment of sausages, fruit syrups and preserves, fresh fruit and hot coffee to round out your menu.

BATTER

1/4 cup butter
1/3 cup sugar
6 eggs, beaten
1 1/2 cups sour cream
1/2 cup orange juice
1 cup flour
2 tsp. baking powder

Combine all ingredients with a mixer or food processor. Butter your largest oven-proof casserole. Make filling.

FILLING
1 pkg. (8 oz.) cream cheese, softened
2 cups small curd cottage cheese
2 egg yolks
1 tbs. sugar
1 tsp. vanilla extract

Combine filling ingredients in a small bowl. Pour half of batter into prepared casserole. Drop filling by the spoonful over batter. Pour remaining batter on top, spreading gently with the back of a spoon. Batter may not cover completely or it mixes with filling. Cover and refrigerate overnight or bake now.

To bake, heat oven to 300°. Bring casserole to room temperature and bake for 35 to 45 minutes until puffed, golden and center is set. Serve with jams, syrups and additional sour cream.

WHOLE LOTTA FRITTATA

Serve with croissants and fresh fruit and your brunch is complete. Sour cream and salsa are wonderful accompaniments.

1 lb. bulk sausage
1 onion, sliced and separated into
   rings
2 zucchini, sliced

$\frac{1}{2}$ lb. mushrooms, sliced
12 eggs
$\frac{1}{2}$ cup dry white wine
salt and pepper

TOPPING

8 oz. sharp cheddar cheese, shredded
2 tomatoes, sliced
1 can (4 oz.) sliced black olives,
   drained

fresh parsley, sour cream and salsa for
   garnish

Heat oven to 350°. Crumble sausage into your largest casserole. Brown sausage for 10 minutes; drain off fat. Add onion, zucchini and mushrooms to mixture and bake for 5 minutes. Beat eggs with wine, salt and pepper. Pour into casserole and continue baking for 12 to 15 minutes, or until eggs are set. Sprinkle with cheese and arrange tomato slices on top. Scatter olives over top. Continue baking for 5 minutes or until cheese is melted and tomatoes are heated through. Let stand for 5 minutes before cutting into wedges to serve. Garnish with parsley, sour cream and salsa.

PUFFED APPLE PANCAKE

This is similar to a Dutch Baby. It's fun to watch it puff in the oven.

1/4 cup butter
2 large apples, peeled and thinly sliced
1/4 cup sugar, divided
1 tsp. cinnamon, divided
6 eggs

1 1/2 cups milk
1 cup flour
1 tsp. vanilla extract
1/2 tsp. salt
3 tbs. brown sugar

Heat oven to 375°. Place butter in an attractive ovenproof casserole and set in oven to melt butter. When butter is melted, add sliced apples and half of the sugar. Sprinkle with half of the cinnamon. Return to oven while making batter. With a mixer or food processor, combine eggs, milk, flour, remaining sugar, cinnamon, vanilla and salt. Pour batter over apples and sprinkle top with brown sugar. Bake for 15 to 20 minutes or until puffed and brown. Serve immediately with hot maple syrup.

BRUNCH EGGS

You can make this up to a day in advance and refrigerate. The recipe also doubles well.

6 eggs, beaten
1 can (17 oz.) cream-style corn
1 can (4 oz.) chopped green chiles, drained
1 cup shredded sharp cheddar cheese
1 cup shredded Monterey Jack cheese
1 tbs. instant grits
1 tsp. Worcestershire sauce
1/4 tsp. white pepper

Heat oven to 300°. Combine all ingredients in a large mixing bowl. Butter a large ovenproof casserole. Pour mixture into prepared pan. Bake for 45 minutes or until firm and light brown. Serve with salsa if desired.

CROISSANT BAKE

If you love croissants as much as we do, you'll find this brunch dish will soon be on your favorites list. The nutmeg enhances the Gruyère, but you could use regular Swiss, Havarti or Monterey Jack cheese as well.

½ lb. bacon, cooked crisp and
  crumbled
4 large croissants, split in half
6 eggs, beaten
1 cup milk
1 tsp. seasoned salt

½ tsp. white pepper
1 tsp. freshly grated nutmeg
2 cups shredded Gruyère cheese
1 cup shredded mozzarella cheese
½ cup shredded Parmesan cheese

Heat oven to 300°. Butter a large ovenproof casserole. Place bottom halves of croissants in dish. Sprinkle with bacon and half of the Gruyère. Combine eggs, milk and seasonings. Pour half of the mixture over croissants. Place top halves on bottoms and sprinkle with remaining Gruyère and mozzarella. Pour remaining custard over croissants and sprinkle with Parmesan. Bake for 30 minutes or until custard is set and cheese is bubbly and brown. If the top gets too brown, cover tightly with foil.

SAUSAGE BREAD

This goes nicely for a crowd when served with scrambled eggs. Broiled grapefruit would be a nice addition—just cut in half, section, drizzle with sherry and brown sugar and bake at 400° until bubbly.

1 1/2 lb. bulk sausage
3 cups biscuit mix
6 eggs, beaten
2 cups milk
1 tsp. salt
1/2 tsp. pepper
1 tsp. dry mustard
1 lb. cheddar cheese, shredded

Heat oven to 300°. Cook sausage in a large skillet over medium heat until no longer pink. Drain off excess fat and set aside to cool. In a large bowl, combine biscuit mix, eggs and milk. Add spices and cooked sausage. Butter two 9-inch cake pans. Divide batter into pans and top with cheese. Bake for 15 to 25 minutes or until bread tests done and cheese is melted. Cut in wedges to serve.

SAUSAGE STRUDEL

This time, flaky pastry is around a savory sausage filling. You can made individual strudels or one large one.

1 lb. bulk sausage
2 tbs. butter
1 tbs. vegetable oil
1 lb. mushrooms, sliced
1 cup chopped onion
1 tsp. salt

1 tsp. pepper
8 oz. cream cheese
12 sheets phyllo
1 cup butter, melted
1 cup fine dry breadcrumbs

Heat oven to 350°. Cook sausage in a skillet over medium heat until no pink remains. Drain and set aside. Melt butter with oil and sauté mushrooms and onion until tender. Sprinkle with salt and pepper. Stir in cream cheese and sausage. Lightly dampen a tea towel. Place a sheet of phyllo on the towel, brush with melted butter and sprinkle with breadcrumbs. Repeat 4 times, ending with 6th sheet of phyllo. Place half of the filling on narrow edge of phyllo, leaving a 2-inch border on each side. Fold in sides and roll up pastry. Place roll on a baking sheet and brush with additional butter. Repeat using remaining phyllo sheets and sausage filling. Bake for 15 to 20 minutes or until golden.

FISH AND SHELLFISH

HALIBUT BAKED IN BUTTERMILK

Buttermilk is a wonderful ingredient. In this recipe, it makes the fish moist and tender as well as brightening the taste.

1 cup buttermilk
salt and white pepper
4 halibut steaks, 4-6 oz. each
1 cup fresh breadcrumbs
1 cup mayonnaise
2 tbs. dry white wine
¼ cup chopped onion
paprika, fresh parsley and lemon slices for garnish

Combine buttermilk, salt and pepper in a shallow dish and marinate fish for 30 minutes. Heat oven to 400°. Remove fish from buttermilk and pat dry with paper towels. Place breadcrumbs in a shallow bowl. Dip fish in crumbs to coat and place in a shallow greased baking pan. Combine mayonnaise, wine and onion. Spread over fish and scatter any remaining breadcrumbs over top. Bake fish for 10 to 15 minutes or until opaque in the thickest part; test with a knife. Sprinkle with paprika and garnish with parsley and lemon.

SESAME HALIBUT WITH LEMON BUTTER

Servings: 4

In the Pacific Northwest, chefs are known for their creative ways with fresh fish. It's fun to purchase wood planks to bake fish; they can be deluxe models purchased at gourmet kitchen stores or simply untreated alder or cedar planks from the lumberyard. The wood plank actually imparts a bit of flavor to the fish. The lemon butter keeps well and is good on vegetables too.

1 cup untoasted sesame seeds
1/2 cup flour
1 tsp. salt
1/4 tsp. white pepper
1/2 cup milk
4 fillets fresh halibut, 4 oz. each
3 tbs. vegetable oil
wooden plank for baking fish, if desired

Heat oven to 400°. In a shallow dish, mix sesame seeds, flour, salt and pepper. Pour milk in another shallow dish. Dip fish in milk and then into sesame seed mixture, turning to coat both sides. In a large skillet over high heat, heat oil to almost smoking. Add fish and brown on both sides, about 1 minute each side. Place fish on a wooden plank or shallow baking dish. Bake for 10 to 12 minutes or until fish feels firm. Remove to serving platter and top each piece with a tablespoon of lemon butter.

LEMON BUTTER
1 cup butter
1 tbs. chopped fresh chives
1 tbs. chopped fresh parsley
1 tbs. chopped garlic
juice and grated zest of 1 lemon

Mix together all ingredients for lemon butter in a medium bowl or food processor workbowl. Using plastic wrap, shape into a log about 2 inches in diameter. Chill or freeze until firm.

HASTY HALIBUT

Steam some fresh green vegetables in foil while the fish cooks, add a tossed green salad and your fresh, fast and healthy dinner is all set.

4 halibut steaks or fillets, 6 oz. each
2/3 cup tartar sauce
1/2 cup finely chopped celery
2 tbs. lemon juice
1/2 cup white wine
1 tsp. paprika

Heat oven to 375°. Place fish in a baking pan and drizzle with lemon juice. Stir celery into tartar sauce. Top each piece of fish with 1/4 of the tartar sauce. Sprinkle tops with paprika. Pour wine in bottom of pan. Bake for 15 to 20 minutes or until fish tests done.

TUNA TERIYAKI

If you have never tasted fresh tuna, you're in for a real treat. It's nothing like the fish you get in cans.

1 cup soy sauce
3 tbs. brown sugar
2 cloves garlic, minced
1 tbs. grated fresh ginger
1/3 cup dry sherry
1 1/2 lb. fresh tuna

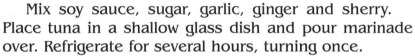

Mix soy sauce, sugar, garlic, ginger and sherry. Place tuna in a shallow glass dish and pour marinade over. Refrigerate for several hours, turning once.

Heat oven to 400°. Remove tuna from marinade and pat dry with paper towels. Spray middle rack with nonstick cooking spray. Place fish directly on rack. Bake for 8 to 12 minutes or until fish tests done with tip of knife inserted in center.

SWORDFISH EN PAPILLOTE

Papillote is a French method of cooking food inside a parchment envelope. Usually meat or fish is partially precooked, placed in the center of a heart-shaped piece of paper and tightly sealed. The enclosed food bakes in its own juices and the paper puffs up and browns. Each guest is given his own papillote to open at the table. It's easy, it's fun, and the aromas are heavenly. Swordfish prepared with the colorful vegetables in this fashion makes a beautiful meal that's also good for you.

1/4 cup flour
salt, pepper and paprika
4 swordfish steaks, about 6 oz. each
2 tbs. vegetable oil
2 tbs. butter
4 cloves garlic, minced
1/4 cup butter
4 green onions, sliced
1 small green bell pepper, cut into matchstick strips
4 oz. mushrooms, sliced
1/2 cup vermouth

Heat oven to 375°. Combine flour, salt, pepper and paprika on a small plate. Dip each side of fish in flour. Melt oil and butter together in a large skillet over medium-high heat. Sauté fish on each side to brown. Remove from heat and set aside. Melt ¼ cup butter and sauté garlic and vegetables until vegetables just begin to soften. Remove from pan. Add vermouth and scrape up any browned bits.

Cut 4 large rectangles of parchment paper. Fold in half crosswise and cut each into a large heart. Place 1 piece of fish on each half of heart and top with vegetables. Spoon pan juices over. Salt and pepper to taste. Fold the edges of each heart tightly to crimp. Place on a baking sheet and bake for 10 to 15 minutes.

BAKED SALMON WITH BASIL BUTTER

Servings: 8

Living in the Pacific Northwest makes fresh salmon a favorite treat. Basil butter is a perfect accompaniment. It's also delectable on cheddar scones.

1 fresh salmon fillet, about 3 lb., or
  8 salmon steaks
1/4 cup butter, melted

1 tbs. lemon juice
2 tsp. Worcestershire sauce
1/2 tsp. garlic powder

Heat oven to 400°. Spray middle rack with nonstick cooking spray. Combine butter with seasonings and brush on both sides of fish. Place fish directly on rack. Bake for 8 to 12 minutes or until fish tests done. Insert tip of a knife in the thickest part. Serve with *Basil Butter*.

BASIL BUTTER

1/2 cup butter
2 tbs. chopped fresh basil
1 tsp. tomato paste

1 clove garlic, minced
freshly ground black pepper

With a mixer or food processor, combine ingredients. Place a piece of plastic wrap on the counter and scrape butter mixture onto plastic. Form into a v-inch log and chill or freeze for several hours before using. This keeps well and can be frozen for several months.

FRESH FISH WITH ALMOND SAUCE

This quick method of preparing fresh fish works well for snapper, sole, black cod, orange roughy or perch. Use either the wine or the lemon juice, but not both.

1 lb. fresh fish fillets or steaks
$\frac{1}{2}$ cup butter, melted
2 tbs. vegetable oil
1 cup sliced almonds
6 green onions, thinly sliced
$\frac{1}{4}$ cup dry white wine, or 2 tbs. fresh lemon juice

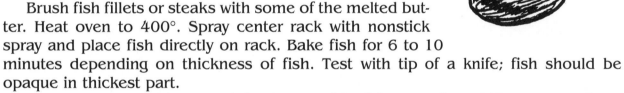

Brush fish fillets or steaks with some of the melted butter. Heat oven to 400°. Spray center rack with nonstick spray and place fish directly on rack. Bake fish for 6 to 10 minutes depending on thickness of fish. Test with tip of a knife; fish should be opaque in thickest part.

Make sauce: Combine remaining butter with oil in a medium skillet over medium-high heat. Add almonds and onions and stir until almonds are beginning to brown and onions are tender. Add wine or lemon juice. Place fish on a platter and pour sauce over top. Garnish with lemon wedges.

PETRALE SOLE WITH PAPAYA AND AVOCADO

Servings: 4-8

This is a glamorous and different fish recipe. Rice Soubise, page 127, would make a nice side dish. Be sure to select papaya and avocados that are not too ripe. If desired, you can garnish fish with finely chopped sliced almonds.

8 petrale sole fillets, rinsed and patted dry
1 papaya, peeled and cut into $1/2$-inch slices
$1/2$ tsp. salt
$1/4$ tsp. white pepper
2 Hass avocados, peeled and cut into $1/2$-inch slices
$1/2$ cup butter
$1/4$ cup chopped fresh parsley
1 tbs. fresh lime juice

Heat oven to 350°. Butter a shallow baking dish. Arrange fish in a single layer on bottom and sprinkle with salt and pepper. Top each piece decoratively with slices of papaya and avocado. Melt half of the butter in a small saucepan and drizzle over fish. Bake for 5 to 8 minutes or until fish flakes. Transfer fish to plates. Heat remaining butter until beginning to brown and add lime juice. Pour over top of fish and fruit, sprinkle with parsley and serve.

RITZY OYSTERS WITH BRIE

Servings: 4-8

This ultra-sophisticated dish can be served for a main course or appetizer.

1 pkg. (10 oz.) frozen chopped spinach, thawed and well drained
1 cup champagne or dry white wine
$1/2$ tsp. dried basil
pinch cayenne pepper
1 qt. shucked oysters, liquid reserved
$1/2$ cup sour cream
$1/2$ cup heavy cream
8 oz. Brie cheese, rind removed and cut into pieces

Place spinach in a shallow quiche or casserole dish. In a saucepan over medium heat, simmer wine, basil and cayenne. Add oysters and cook until plump and firm, about 3 minutes.

Heat oven to 300°. With a slotted spoon, place oysters on top of spinach. Place in oven. Bring wine mixture to a boil. Cook until liquid is reduced to $1/4$ cup. Add sour cream and heavy cream and cook until bubbly. Whisk in Brie, continuing until it melts. Pour over oysters and sprinkle with Parmesan. Bake for an additional 5 minutes. Serve on dinner plates as a main course, or on small plates as an appetizer.

BACON-WRAPPED SCALLOPS WITH VODKA MAYONNAISE

These can be served as a main dish or an appetizer (serves 8 as an appetizer). Soak wooden skewers in water for an hour before using so they won't burn. A pureed avocado or tomato paste are also fun to add to the vodka mayonnaise.

1 lb. bacon
1 lb. large scallops
wooden skewers
1 cup mayonnaise
zest of 1 lemon
2 tbs. vodka

Heat oven to 400°. Cut bacon into thirds, place in a pan of boiling water and blanch for 5 minutes. Place on paper towels to drain and cool. Wrap each scallop with a piece of bacon and thread onto skewers. Place skewers on center rack and cook for 8 to 12 minutes or until scallops are opaque and bacon is crisp. Combine mayonnaise with lemon zest and vodka. Serve scallops hot with sauce for dipping.

GREEK SHRIMP

Serve this quick and elegant dinner entrée with a Greek salad and orzo, rice-shaped pasta.

2 tbs. olive oil
2 cloves garlic, minced
1 lb. jumbo shrimp, peeled and deveined
3 tomatoes, peeled, seeded and coarsely chopped
1/2 tsp. crumbled dried oregano
4 oz. feta cheese, crumbled

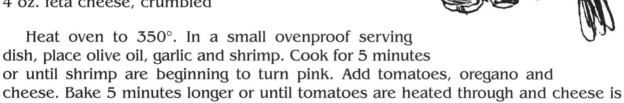

Heat oven to 350°. In a small ovenproof serving dish, place olive oil, garlic and shrimp. Cook for 5 minutes or until shrimp are beginning to turn pink. Add tomatoes, oregano and cheese. Bake 5 minutes longer or until tomatoes are heated through and cheese is melted. Do not overcook or shrimp will toughen.

SCAMPI

Simple and delicious. Prepare it completely in advance and then pop it in the oven just 10 minutes before serving. Garlic put through a garlic press has more flavor and aroma than that which is simply crushed or chopped. The Zyliss Swiss-made garlic press is an excellent kitchen tool, and I wouldn't be without one.

1/2 cup butter, melted
1/4 cup olive oil
4 cloves garlic, minced
4 shallots, minced
3 tbs. finely chopped fresh parsley
3 tbs. lemon juice
salt and pepper
1 1/2 lb. scampi or large shrimp, peeled and deveined

Combine butter, oil, garlic, shallots, parsley, lemon juice and seasonings in a shallow ovenproof casserole. Add scampi and toss to coat well. Can be covered and refrigerated at this time. To cook, heat oven to 400°. Place dish on center rack and cook for 8 to 10 minutes. Stir once to evenly distribute sauce.

CRAB CROISSANTS

This is perfect for lunch when you want something very special.

$1/2$ cup mayonnaise
1 clove garlic, minced
$1/4$ tsp. dried dill
dash cayenne pepper
1 tbs. finely chopped fresh parsley
1 green onion, finely chopped
$1/2$ lb. crabmeat
1 can (4 oz) sliced ripe olives, drained
2 artichoke hearts, coarsely chopped
4 oz. cheddar cheese, shredded
3 croissants, split in half horizontally

Heat oven to 325°. In a bowl, combine mayonnaise, garlic, dill, cayenne, parsley and onion. Stir in crab, olives, artichokes and $3/4$ of the cheese. Place croissants, cut side up, on a baking sheet. Spread crab mixture on croissants and top with remaining cheese. Bake for 8 to 12 minutes or until bubbly and crab is heated through.

MEATS

PORK TENDERLOIN IN PUFF PASTRY

Servings: 2-4

This technique is quite easy and the results most impressive. The recipe can easily be doubled. Serve with a leaf lettuce salad with Mandarin oranges, sliced almonds and a sweet and sour dressing.

1 pork tenderloin
salt and pepper
¼ cup Dijon mustard
¼ cup brown sugar, packed

¼ cup butter
1 sheet puff pastry
1 egg
pinch salt

Heat oven to 350°. Trim pork tenderloin so it is of even thickness: either trim small end off or fold under. Sprinkle meat with salt and pepper and spread with mustard. Pat brown sugar over surface to coat. In a large skillet, heat butter until melted. Sauté pork on all sides until it is evenly browned and sugar has melted to form a coating, about 5 minutes total. Place pork in center of puff pastry sheet and fold to enclose completely. Pinch seam to seal. Trim any excess and set aside to decorate. Place pork seam-side down on a greased baking sheet. Mix egg with a bit of salt and brush pastry with egg glaze. Use scraps of pastry to make leaves, shapes or other decorative designs. Brush again with egg glaze. Bake for 20 to 30 minutes or until nicely puffed and golden. Cut into slices to serve.

CROWN ROAST OF PORK
WITH WILD RICE STUFFING

This is a spectacular holiday entrée. Use Fruited Rice, *page 125 or* Wild and White Rice Pilaf, *page 128. Paper frills or kumquats look pretty on the tips of the bones. Order your roast ahead of time from the butcher and allow one or two ribs per person.*

1 tbs. salt
1 tsp. black pepper
1 tsp. dried sage
1 tsp. dried thyme
2 bay leaves, crumbled

$1/4$ tsp. allspice
3 cloves garlic, minced
1 crown roast of pork (8 or 16 ribs)
$1/4$ cup vegetable oil

Combine salt, spices and garlic. Rub into meat. Place in a large plastic bag and refrigerate 24 hours. Turn bag several times to distribute juices.

Heat oven to 300°. Rub marinade off meat and pat dry with paper towels. Add oil to roasting pan. Place meat in center, bones up. Roast for 15 minutes per pound. Meat thermometer should register 180° when inserted into thickest part of roast. Twenty minutes before roast is done, mound stuffing in the center. If bone tips get too brown, cover tightly with foil or put frills or kumquats on ends after roasting.

LUAU PORK

It's such fun to have your own summer luau, and this recipe is the foundation for a wonderful feast. Add an assortment of tropical fruits for a salad, some Mai Tais and Macadamia Nut Dessert, page 142, and you're all set. If you have any meat left over, it makes a terrific sandwich on egg bread with pineapple cream cheese and a slice of raw Maui onion.

12 ti leaves, optional (available from
    your florist)
5 lb. boneless pork butt
¼ cup soy sauce
1 tsp. Worcestershire sauce

3 cloves garlic, minced
one 3-inch piece fresh ginger, sliced
1 tbs. salt
1 tbs. liquid smoke

Heat oven to 300°. Place a large sheet of heavy duty foil in a baking pan. Place 3 ti leaves on the foil. Put roast on top and sprinkle with remaining ingredients. Place remaining leaves on top of pork and seal foil very tightly. You might wish to wrap it a second time. Place pan on lower rack of oven and bake for 3 to 4 hours. Meat should fall apart and be very tender and juicy. To serve, shred pork and mound on a platter. Garnish with leaves and flowers.

PLUM-SAUCED SPARERIBS

Servings: 4

We like to use boneless country-style ribs for this recipe. This sauce is also good on chicken wings as an appetizer.

4 lb. country-style spareribs
1 tsp. salt

$\frac{1}{2}$ tsp. pepper
2 cloves garlic, minced

Heat oven to 325°. Place ribs in a shallow baking pan and bake for 1 hour. Drain off fat.

SAUCE

1 cup plum jam
$\frac{1}{3}$ cup dark corn syrup
$\frac{1}{3}$ cup soy sauce

2 cloves garlic, minced
3 green onions, finely chopped
1 tbs. finely chopped fresh ginger

Combine sauce ingredients. Pour sauce over ribs and continue baking for 1 hour or until fork-tender.

FAVORITE MEAT LOAF

Servings: 8

Kids particularly like the glaze on this meat loaf. Experiment by adding other chopped vegetables, such as celery, green pepper or carrot.

2 lb. lean ground beef
1 medium onion, chopped
2 slices whole wheat bread
1 egg
1/4 cup minced fresh parsley

1 tsp. seasoning salt
1/2 tsp. pepper
1/2 tsp. nutmeg
1/2 cup V-8 juice or catsup

Heat oven to 300°. In a large bowl, combine all ingredients until evenly mixed. Pat into a loaf pan or, if you prefer, shape into a large flat round, which will result in a crisper surface. Place the round on a pie pan or a shallow casserole. Bake for 45 minutes or until center tests done. Prepare glaze.

GLAZE

1/2 cup catsup
2 tsp. dry mustard

2 tbs. dark brown sugar
dash Tabasco Sauce

Combine ingredients in a small bowl and pour over top of meat loaf. Serve hot or use cold for sandwiches.

BLACKENED TENDERLOIN
WITH JALAPEÑO HOLLANDAISE

It's easy to make your own "blackened" seasoning and to adjust the heat to suit your personal taste. Try the mixture on rib eye steaks, chicken breasts or any firm fish such as halibut or salmon.

SEASONING MIXTURE

1 tbs. paprika
2½ tsp. salt
1 tsp. onion powder
1 tsp. garlic powder
1 tsp. cayenne pepper

¾ tsp. white pepper
¾ tsp. black pepper
½ tsp. dried thyme
½ tsp. dried oregano

Combine all seasonings and place in an airtight container. Store at room temperature.

4 beef tenderloins, 4-6 oz. each
$\frac{1}{4}$ cup butter, melted
seasoning mixture

Heat oven to 400°. Dip tenderloins in butter and then in seasonings. Bake on center rack for 7 to 10 minutes, depending on desired doneness and thickness of meat.

JALAPEÑO HOLLANDAISE SAUCE
4 egg yolks
dash Tabasco Sauce
2 tbs. jalapeño juice
1 tbs. chopped jalapeño chiles
$\frac{1}{2}$ cup butter, melted and hot

With a food processor or blender, combine egg yolks, Tabasco, juice and chopped jalapeños. With motor running, slowly drizzle butter into feed tube. Serve immediately or keep warm in a vacuum bottle.

MARINATED FLANK STEAK
WITH GREEN PEPPERCORNS

Servings: 4

Marinate your steak up to 24 hours in advance for best flavor.

1 flank steak, 2 lb.
3 tbs. butter, softened
2 tbs. Dijon mustard
1 tsp. Worcestershire sauce
1/2 tsp. salt
1/2 tsp. pepper

1/2 cup burgundy wine
1 cup sour cream
3 tbs. brandy
1 tbs. green peppercorns in brine,
  rinsed

Place steak in a shallow dish. Make a paste of butter, mustard, Worcestershire, salt and pepper. Spread on steak. Pour wine over steak. Cover and refrigerate for 24 hours. Turn occasionally.

Heat oven to 400°. Drain meat and pat dry. Reserve marinade. Cook steak for 6 to 8 minutes or until medium rare. Place reserved marinade in a small saucepan and heat. Whisk in sour cream, brandy and green peppercorns. Do not boil. Slice steak on the diagonal and serve with sauce.

RACK OF LAMB

Now you can prepare this special restaurant dish at home.

2 racks of lamb
1 clove garlic, minced
1/2 tsp. salt

1/2 tsp. dried thyme
3 tbs. Dijon mustard
3 tbs. olive oil

CRUMB COATING

6 tbs. butter
3/4 cups breadcrumbs, recommend
   sourdough

1 clove garlic, minced
1/4 cup chopped fresh parsley
salt and pepper

   Heat oven to 400°. Score racks of lamb on top. Combine garlic, seasonings, mustard and oil and spread on lamb. Place lamb in a shallow baking dish and bake for 10 minutes. In a skillet, melt butter and toss bread crumbs with garlic, parsley, salt and pepper until toasted. Remove lamb from oven and pat crumbs on surface. Continue to roast lamb for 10 to 15 minutes until crumbs are golden and meat tests 140° with a meat thermometer for rare, 160° for medium rare and 170° for well done.

BUTTERFLIED LEG OF LAMB

Servings: 8

This marinade is fabulous! We trim the meat fastidiously when we get it home to get rid of any excess fat or gristle. Cubing the lamb for shish kebob works well too. Couscous makes a delicious side dish. Cook it in chicken broth and add sliced green onions, dried currants, toasted walnuts, garlic chili oil and lots of chopped parsley.

1 leg of lamb, butterflied, about 4-5 lb.	1 tsp. dried basil,
2 cloves garlic, minced	1 tsp. dried oregano
1/4 cup lemon juice	1 tsp. dried thyme
1/2 cup red wine vinegar	1 tsp. dried rosemary
3/4 cup vegetable oil	1 1/2 tsp. salt
1/2 cup Dijon mustard	1/2 tsp. freshly ground black pepper

Place lamb in a shallow glass container or a large plastic bag. Combine remaining ingredients to make a marinade, and pour over lamb. Marinate for at least 24 hours or longer.

To cook, heat oven to 400°. Remove meat from marinade and pat dry with paper towels. Place lamb on center rack and immediately turn heat to 300°. Roast lamb for about 20 minutes per pound or until internal temperature reaches 175° for well or 165° for slightly rare.

POULTRY

CRAB-STUFFED CHICKEN BREASTS

Servings: 8

This is an elegant entrée for a dinner party. Serve with a green salad, rice, crusty French bread and a dry white wine.

4 large whole chicken breasts, halved, skinned and boned
1/3 cup sliced green onions
1/3 cup sliced mushrooms
1/4 cup butter
3 tbs. flour
1/2 tsp. dried thyme
3/4 cup chicken stock
1/2 cup milk
1/3 cup dry white wine
salt and white pepper
1 cup shredded Swiss cheese
1 cup fresh crabmeat
2 tbs. minced fresh parsley
1/3 cup dry breadcrumbs
1/2 cup sliced almonds

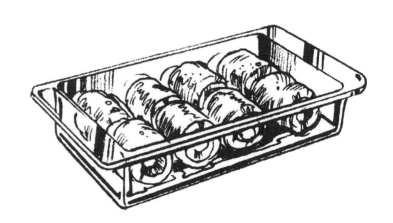

Heat oven to 325°. Pound chicken breasts 1/4 inch thick.

In a skillet, sauté onions and mushrooms in butter until tender. Sprinkle with flour and stir to combine. Add thyme, chicken stock, milk and wine to mixture and cook until smooth. Add salt and pepper to taste. Stir in half of the Swiss cheese and cook until sauce is thickened.

In a small bowl, stir together crab, parsley and breadcrumbs. Add 1/4 cup sauce to mixture and combine well. Divide mixture into eighths and place in center of each chicken breast. Roll meat around stuffing and place seam-side down in a buttered baking dish. Pour remaining sauce over chicken.

Bake for 20 minutes, sprinkle with remaining cheese and sliced almonds, and continue cooking 5 minutes longer, until cheese is melted and almonds are brown.

CHICKEN BOURSIN IN RED WINE SAUCE

Servings: 8

As you bite into this delectable dish you encounter tender chicken, slightly salty prosciutto ham, garlicky herb cheese and a rich, robust wine sauce. Serve a bulgur pilaf, marinated mushroom salad and baby carrots to complete your menu.

4 whole chicken breasts, halved and boned
3 tbs. white wine
salt and pepper
8 oz. Boursin cheese or other herbed cream cheese
8 thin slices prosciutto ham

Heat oven to 400°. Slightly flatten each breast, removing skin if desired. Sprinkle each breast with white wine, salt and pepper. Butter a small baking dish. Divide cheese into 8 equal portions. Wrap each portion in a slice of the ham, enclosing cheese completely. Place in center of each chicken breast, shaping the meat around the cheese and ham to form a compact bundle. Place in prepared dish.

Bake for 15 to 20 minutes or until chicken is cooked through; test by making a small slit with the tip of a knife. Chicken should be completely white.

RED WINE SAUCE

3 tbs. butter
½ cup minced shallots
½ cup minced mushrooms
3 tbs. flour
1 can (10 oz.) beef consommé, undiluted
½ cup robust red wine
1 tbs. Bovril beef extract, optional
salt and cayenne pepper

In a skillet, melt butter and sauté shallots and mushrooms until soft. Sprinkle with flour and stir to combine. Add consommé, wine, beef extract and seasonings; cook until bubbly. Taste and correct seasoning. Keep sauce warm over low heat and serve with chicken.

ROAST CHICKEN STUFFED WITH CHEESE

Servings: 4

The aroma of this dish cooking is heavenly, as it turns a lovely golden brown.

1 large whole fryer, about 4 lb.
2 cloves garlic, minced
3 tbs. chopped fresh parsley
3 tbs. chopped chives or green onions
1 tbs. lemon juice

1/2 cup butter, room temperature
4 oz. cream cheese
4 oz. blue cheese
salt and freshly ground black pepper
paprika

Heat oven to 375°. Rinse chicken and pat dry. Butterfly chicken by cutting through backbone on each side from tail to neck. Remove backbone and discard. Turn chicken over and flatten by cracking breast bones. Trim off any excess fat or skin and cut off wing tips. Gently loosen skin, starting at breast end, by inserting your hand between flesh and skin. Be careful not to pierce skin. Loosen flesh all the way through to legs, leaving skin at edges of chicken intact.

Combine remaining ingredients, except salt, pepper and paprika. Distribute cheese mixture evenly under skin and work it all over bird with your fingers. Place chicken in a pan skin-side up. Sprinkle with salt, pepper and paprika.

Bake for 35 to 50 minutes or until chicken is puffed and brown and juices run clear when breast is pierced with a knife. To serve, cut in quarters with kitchen shears. Spoon juices over portions.

POCKET CHICKEN BREASTS

Servings: 4-8

These are good hot or at room temperature, which makes them great for picnics or outdoor entertaining. The stuffing is particularly colorful and flavorful and helps to keep the breast moist and tender. We like to serve it with a fruited wild rice salad.

4 whole chicken breasts with skin
1/2 lb. bacon
1 large onion, chopped
1 pkg. (10 oz.) frozen chopped
   spinach, thawed
1 egg

1/4 cup freshly grated Parmesan cheese
2 tbs. chopped pimientos or roasted
   red pepper
salt and pepper
2 tbs. butter, melted

Heat oven to 325°. Remove bones from chicken and leave skin intact. Cut each breast in half and set aside. Fry bacon until crisp. Crumble and set aside. Remove all but 2 tbs. bacon drippings from pan. Sauté onion until tender. Squeeze moisture from thawed spinach. Combine bacon, onion, spinach, egg, cheese, pimiento and seasonings. Lift skin from each breast half and place 1/8 of the mixture on top of the breast. Fold edges of chicken skin underneath to make a compact bundle.

Place breasts in a buttered pan and brush with 2 tbs. of melted butter. Bake for 20 to 30 minutes or until done, checking with point of a knife. Serve hot, warm or at room temperature.

BACON-WRAPPED CHICKEN

In this delectable dinner dish, the bacon bastes the chicken as well as adding flavor. You could vary the filling to suit your personal taste — spinach and sun-dried tomatoes, broccoli and cheddar or mushrooms and ham would also be nice combined with the cream cheese.

8 oz. cream cheese, room temperature
2 oz. blue cheese crumbles
1 clove garlic, minced
1 tbs. minced fresh parsley
1/2 tsp. salt
4 whole chicken breasts, halved, skinned and boned
8 slices bacon

Combine cream cheese, blue cheese and seasonings. Spread each breast half with cheese mixture. Roll up each breast starting at the small end. Wrap a slice of bacon around each bundle, overlapping edges slightly. Place seam side down in a baking dish. You can cover and refrigerate the dish at this point.

To bake, heat oven to 325° and cook chicken for 15 to 20 minutes or until bacon is crisp and breast is cooked through; test with the tip of a knife.

CHICKEN PEKING

Fried rice, fresh asparagus and a fruit salad go nicely with this succulent chicken.

1 chicken, 4 lb., or equivalent weight in chicken pieces
$1/2$ cup soy sauce
$1/4$ cup dry sherry
$1/3$ cup hoisin sauce
6 green onions, minced
2 cloves garlic, minced

$1/4$ cup cider vinegar
$1/2$ cup honey
$1/4$ cup orange marmalade
1 tsp. grated orange zest
1 tsp. grated lemon zest
2 tbs. minced green bell pepper
4 drops Tabasco Sauce

Place chicken in a glass dish. Combine remaining ingredients in a saucepan and bring to a boil over medium heat, stirring. Lower heat and simmer 5 minutes. Cool marinade. Pour over chicken and marinate in refrigerator overnight.

Heat oven to 325°. Drain chicken and reserve marinade. Place chicken in a shallow greased baking dish and bake for 30 to 40 minutes, basting occasionally with reserved marinade. Chicken will be richly glazed. Check for doneness by inserting the tip of a knife into the thickest part of the breast. Juices should run clear and meat should be cooked through.

CHICKEN IN PHYLLO

Although these are time-consuming to prepare, they can be made ahead and frozen and baked at your leisure. Prepare for the compliments.

6 whole chicken breasts, skinned, halved and deboned
1 pkg. (10 oz.) frozen chopped spinach, thawed and squeezed dry
1 cup shredded Swiss cheese
1/4 cup chopped walnuts
1/2 cup ricotta cheese

1 tbs. green peppercorns, in brine, rinsed
1/2 tsp. salt
1/4 tsp. pepper
1 pkg. phyllo
1 cup butter, melted

Pound chicken breasts flat. In a bowl, combine spinach, Swiss cheese, walnuts, ricotta, green peppercorns and seasoning; mix well. Spread mixture evenly on breasts and roll up to form a bundle. Place each breast in the center of the short end of a sheet of phyllo. Fold 1/3 over and brush with butter. Repeat with other side and brush lightly with butter. Roll chicken to end and place on a baking sheet. Brush top with butter. Repeat with remaining breasts and phyllo.

Heat oven to 350° and bake for 20 to 30 minutes or until chicken breasts are cooked through and pastry is golden.

ROSEMARY GAME HENS

Depending on appetites and accompaniments, serve a whole or a half game hen per person. Cut in half after baking.

4 Cornish game hens, about 1½ lb. each
1 tsp. salt
½ tsp. pepper
¼ cup butter, melted
4 cloves garlic, minced
2 tbs. fresh rosemary
1 lemon, quartered

Heat oven to 375°. Wash hens and pat dry.
Place in a shallow baking dish, breast side up.
Sprinkle with salt and pepper. Combine butter, garlic and rosemary. Brush on hens.
Place a lemon quarter in each cavity. Bake hens for 20 to 30 minutes, basting with butter mixture.

CRISP ROASTED DUCKLING

Whenever we want to impress guests, this is one of our favorites. Be sure to read the instructions carefully. Air-drying the duck for several days results in delightfully crisp skin. Be sure to start at least 4 days in advance, because if your duckling is frozen, you will need 2 days to thaw it in your refrigerator. The Peach Chutney makes a nice gift from your kitchen.

1 duckling, thawed, 5-6 lb.
salt and pepper

Rinse thawed duckling and discard giblets. Pat dry with paper towels. Trim any excess fat off neck and tail. Place duck in a shallow dish in the refrigerator uncovered for 2 days. Skin should shrink and dry.

Heat oven to 400°. Line a baking pan with foil. Prick duckling all over with a fork. Spray foil with nonstick cooking spray. Place duckling breast side down in oven, and sprinkle with salt and pepper. Bake for 30 minutes. Remove pan from oven and pour off fat. Turn duck breast side up and roast for an additional 20 to 25 minutes. Remove duckling from pan and cool to room temperature.

Using scissors, cut duckling down back on one side of backbone. Repeat on other side and discard backbone. Cut duckling in half lengthwise through breastbone. Cut duck into quarters just above thigh. You will now have 4 nice pieces of semi-boned duckling. Just before serving, re-crisp duckling. Heat oven to 400° and place duck pieces skin side up in a shallow roasting pan. Bake for 12 to 18 minutes, until piping hot and skin is crisp. Serve with *Peach Chutney*.

PEACH CHUTNEY

Makes 5 cups

1 can (29 oz.) sliced peaches
1 onion, chopped
2 cups cider vinegar
2 cups brown sugar, packed
1 clove garlic, minced
1/2 cup chopped crystallized ginger

2 tbs. mustard seed
2 tsp. chili powder
1 tsp. ground cloves
1 tsp. salt
1 cup dark raisins

Drain peaches and set fruit aside, retaining syrup. In a large saucepan, combine peach syrup with remaining ingredients. Bring to a boil and cook for 30 minutes over medium low heat, stirring occasionally. Chop peaches coarsely and add to mixture. Continue cooking until thickened, about 30 minutes. Pour into jars and refrigerate. Keeps well for 1 month. Freeze for longer storage.

VEGETABLES

SHERRIED TOMATOES

Tasty and colorful, these tomatoes are easy to make and even easier to eat. We've used tomatoes from the store that were marginal at best and they still tasted great.

8 medium tomatoes
1/2 cup sherry
dill and black pepper
1 cup mayonnaise
1 cup shredded cheddar cheese

Heat oven to 400°. Cut top portion from each tomato. Turn upside down over the kitchen sink or a bowl and gently squeeze to remove seeds and juice. Pierce each tomato with a fork and sprinkle with sherry, dill and black pepper. Place tomatoes in a shallow baking dish and bake for 5 to 8 minutes or until heated through and beginning to soften. Combine mayonnaise and cheddar. Place a heaping spoonful on top of each tomato. Increase oven heat to 500° and continue cooking until top is puffed and golden.

AUTUMN VEGETABLE MELANGE

Servings: 4-6

If you don't love parsnips, now is a good time to start. Their sweet flavor really enhances this colorful combination of fall vegetables.

3 carrots, peeled and sliced
2 small zucchini, sliced
2 summer squash, sliced
2 parsnips, peeled and sliced
$1/4$ cup butter
salt and pepper
2 tbs. water

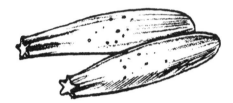

Place prepared vegetables on a large sheet of foil. Dot with butter and sprinkle with salt and pepper. Sprinkle vegetables with 2 tbs. water. Seal foil tightly. Place vegetables on a rack in the oven and cook with the rest of your meal, about 20 to 30 minutes, depending on the temperature you are using.

CARROTS WITH GINGER AND LEEKS

This is a nice combination to go with chicken or pork.

8 carrots, peeled and cut into matchstick strips
2 leeks, white part only, cut into matchstick strips
1/3 cup butter
2 tbs. minced fresh ginger
1 tsp. sugar
1 tsp. salt
1/2 tsp. white pepper

Heat oven to 325°. Place vegetables on a large piece of foil. Cut butter into slices and place on top. Scatter ginger over the top and sprinkle with sugar and seasonings. Fold foil packet up tightly to seal. Bake for 30 minutes, or until vegetables are tender-crisp.

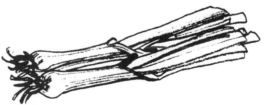

OVEN-ROASTED BABY CARROTS
WITH HONEY GLAZE

An excellent side dish with chicken or pork, this makes ordinary carrots something special.

2 lb. baby carrots
1/3 cup butter
2 large shallots, finely chopped
1 tbs. fresh thyme
1 tsp. salt
1/2 tsp. white pepper
1 cup honey
1/4 cup dried apricot strips
1/2 cup sweet white wine, such as Riesling

Heat oven to 325°. Spread carrots on a rimmed baking sheet and roast for 30 minutes, or until tender. Cover apricots with wine and microwave for 1 minute on high setting.

To make glaze, melt butter in a large saucepan and cook shallots until transparent. Add honey, thyme and apricots. Cook until mixture is reduced by one third. Add carrots and toss to coat. Serve hot.

MINTED PEAS IN TOMATO CUPS

Servings: 8

These make a tasty and colorful garnish.

4 large tomatoes, cut in half, centers removed
1 pkg. (10 oz.) frozen tiny peas
1/2 cup mint sauce (prefer Cross and Blackwell) or mint jelly

Hey oven to 325°. Place tomatoes in a pretty shallow ovenproof casserole. Thaw peas and mix with mint sauce or jelly. Fill tomato cups. Bake for 20 to 30 minutes or until piping hot.

OVEN-ROASTED ONIONS

These are excellent served with steak, roasts or chicken. They taste rich, yet are low in calories. Puree any leftovers with butter and a tablespoon each of red wine and balsamic vinegar, and serve with French bread as an appetizer or toss with pasta.

12 small onions, about 2 inches in diameter
2 tbs. olive oil
2 tbs. balsamic vinegar
salt and freshly ground pepper

Heat oven to 375°. Place onions in a baking dish and drizzle with oil and vinegar. Bake onions for 1 hour or until richly browned. Turn them over occasionally.

ONION FLOWERS WITH TOASTED PECANS

These beautiful onions look very dramatic on a holiday buffet. We particularly like them with Crown Roast of Pork, page 86.

12 small onions, about 2-inch diameter
6 tbs. butter, melted
salt and freshly ground pepper
3 tbs. chopped pecans

Heat oven to 350°. Make flowers out of onions by making multiple slits in each onion ¾ of the way through to the root end. Place onions root-end down in a baking pan. Drizzle with melted butter, spread the "petals" out and sprinkle with salt and pepper. Bake for about 1 hour. Add pecans to center of onions and bake for another 20 minutes.

ONIONS FILLED WITH BROCCOLI

For a holiday meal, garnish these with a bit of pimiento.

1 pkg. (10 oz.) frozen broccoli
4 large Spanish onions
3/4 cup mayonnaise
1/2 cup grated Parmesan cheese
2 tsp. lemon juice
1 tsp. salt
1/2 tsp. pepper

Cook broccoli until tender-crisp and drain well. Peel onions and cut in half cross-wise. Gently parboil onions in salted water for 10 minutes. Drain well. Remove center of each onion half, leaving a 3/4-inch wall.

With a food processor, combine onion centers, broccoli, mayonnaise, cheese, lemon juice and seasonings. Mound mixture into onion shells. May be prepared in advance to this point and refrigerated. To bake, heat oven to 325° and place onions in a buttered ovenproof casserole. Bake for 15 to 20 minutes or until broccoli is bubbly and onions are beginning to brown.

BAKED PARSNIPS WITH APPLES AND ORANGE

Servings: 8

Something a bit different — bake some stuffed Cornish game hens at the same time for a lovely autumn dinner.

2 oranges
1 lb. parsnips, peeled and thinly sliced
2 Granny Smith apples, peeled, cored and thinly
   sliced
1/2 tsp. salt
1 tbs. brown sugar
1/4 cup butter, melted

Heat oven to 350°. Juice one of the oranges. Remove peel and all white membrane from remaining orange and cut into thin slices. On a double sheet of foil, place parsnips, apples and orange slices. Sprinkle with salt and brown sugar. Drizzle with melted butter and orange juice. Tightly seal foil and bake for 1 hour.

SPINACH PIQUANT

This wonderful combination of flavors goes well with poultry, pork or fish.

2 jars (6 oz. each) marinated artichokes, drained
2 pkg. (10 oz.) frozen chopped spinach, thawed
8 oz. cream cheese, softened
2 tbs. butter
1/2 cup grated Parmesan cheese
1 tsp. paprika

Heat oven to 350°. Place drained artichokes in the bottom of an ovenproof serving dish. Squeeze excess moisture out of thawed spinach. Layer spinach over artichokes. In a small bowl, combine cream cheese, butter and half of the Parmesan cheese. Spread over spinach. Sprinkle with remaining cheese and paprika. Bake for 15 to 25 minutes or until heated through and top is bubbly.

BRUSSELS SPROUTS WITH BROWNED BUTTER

Servings: 6

Browned butter enhances any green vegetable. The French call it beurre noisette and sometimes add a bit of lemon juice as well. Try this technique with asparagus, cauliflower, broccoli or green beans—wonderful.

1 lb. Brussels sprouts
$1/2$ cup butter
salt and freshly ground pepper

Heat oven to 400°. Wash and trim Brussels sprouts. Place sprouts on a large square of foil and sprinkle with 1 tbs. water. Seal pouch tightly and place on center rack. Place butter in an ovenproof casserole. Bake pouch and butter for 5 to 8 minutes or until butter is melted and beginning to turn nut brown. Remove butter and set aside. Continue baking sprouts for 10 minutes longer.

Open pouch carefully and place sprouts in a warm serving dish. Drizzle with browned butter and sprinkle with salt and pepper.

ROASTED GARLIC

What wonderful things roasting does to whole heads of garlic! The flavor is rich and mellow. A popular restaurant in the Seattle area serves an appetizer of cloves of roasted garlic, thinly sliced Hass avocado and goat cheese, all drizzled with olive oil on an ovenproof plate. It is served warm with crisp thin crostini for spreading. Crostini are the equivalent of Italian melba toast.

4 whole heads garlic
1 tbs. olive oil
salt and pepper to taste

Heat oven to 300°. Slice top 1/2 inch from garlic. Remove some of the papery outer skin. Keep heads intact. Place in an ovenproof shallow casserole, such as a glass pie plate. Drizzle with oil and sprinkle with salt and pepper. Bake for 1 hour or until tender. To serve, squeeze warm cloves out of their skins onto crusty bread or crackers and spread with a knife.

NUTTY SQUASH

This is a good way to get children to eat squash.

2 whole acorn squash
1 cup cracker crumbs or graham cracker crumbs
1 cup chopped pecans
1/2 cup butter, softened
1/2 cup brown sugar, packed
1/2 tsp. salt
1/2 tsp. nutmeg

Leave squash whole and pierce several times with a sharp knife. Microwave on high for 15 to 20 minutes or until tender. Cut in half and remove seeds. Place halves cut-side up in a baking dish. Heat oven to 300°. In a medium bowl, combine crumbs, nuts, butter, sugar and seasonings until crumbly. Divide equally among squash halves. Bake for 12 to 15 minutes or until golden.

MUSHROOM CASSEROLE

Make this a day ahead and bake on the center rack while your roast, chicken or chops bake on the bottom.

1/4 cup butter
1 1/4 lb. mushrooms, sliced
1 small onion, chopped
3 stalks celery, sliced
1 green bell pepper, chopped
6 slices bread, crust removed and
    cubed

2 eggs
1/2 cup mayonnaise
1 1/2 cups milk
1 tsp. salt
1/2 tsp. pepper
1/4 cup shredded Parmesan, Swiss or
    cheddar cheese

Melt butter in a large skillet over medium high heat. Sauté mushrooms, onion, celery and green pepper until soft. Butter an ovenproof casserole. Place half of the bread in the bottom. Pour mushroom mixture over. Top with remaining bread cubes. Beat eggs, mayonnaise, milk and seasonings. Pour over top. Cover and refrigerate overnight. Heat oven to 325°. Bake for 40 minutes to 1 hour. Sprinkle with cheese before the last 10 minutes of baking.

SIDE DISHES

MEXICAN RICE WITH BLACK BEANS

Black beans are sometimes called "turtle beans." They can be purchased at health food stores or canned. To cook, soak overnight, drain and cover with fresh water. Bring to a boil, reduce heat and simmer until tender, about 30 minutes. To use the canned variety, simply drain and rinse with cold water. This is a nice dish for a potluck for vegetarian friends.

6 cups cooked rice
2 cups cooked black beans
1/2 lb. ricotta cheese
1/4 cup milk
6 oz. Monterey Jack cheese, shredded

3 cloves garlic, minced
1 large red onion, diced
1 can (4 oz.) diced green chiles
4 oz. cheddar cheese, shredded

Heat oven to 325°. Mix rice and beans together in a large bowl. Whisk together ricotta, milk and remaining ingredients, except cheddar cheese. Combine with rice and beans and place in a greased casserole. Bake for 20 minutes. Top with shredded cheddar and continue baking until melted, about 5 minutes.

FRUITED RICE

This is a wonderful side dish to serve in the autumn with pork or chicken.

2 tbs. butter
2 cups thinly sliced carrots
1 1/4 cups water
3/4 cup apple juice
1 cup rice
1 tsp. salt

1/2 tsp. cinnamon
2 red apples, cored and cubed
2 tbs. lemon juice
1/2 cup raisins
1/2 cup green onions
1 tbs. sesame seeds, toasted

Heat oven to 300°. In an ovenproof 4-quart saucepan or casserole with a lid, melt butter and cook carrots for 5 minutes. Add water, apple juice, rice, salt and cinnamon. Cover and cook for 15 minutes. Toss apples with lemon juice and add to pan with raisins, onions and sesame seeds. Cook 5 minutes longer. Fluff with a fork and serve.

CHEESY RICE

This recipe is ideal for buffets. It can be made the day before and refrigerated. This basic concept can be varied to suit your individual taste: try a different combination of cheeses, add diced red peppers, bacon, or top with Parmesan — the possibilities are endless.

3 cups cooked white rice
8 oz. cheese, shredded: Monterey Jack or cheddar or a combination
1 cup finely chopped fresh parsley
1 cup finely chopped green onions
1 can (4 oz.) diced green chiles
2 eggs
1 cup milk
1/3 cup butter, melted
salt and pepper

Heat oven to 325°. Combine all ingredients and place in a buttered casserole. Bake for 25 to 35 minutes, or until set like a custard.

RICE SOUBISE

Elegant, simple but sophisticated, this rice goes well with fish or chicken. The bay leaf and vermouth add so much flavor. The French word "soubise" refers to onions in a recipe.

$\frac{1}{2}$ cup butter
2 cups finely minced onions
1$\frac{1}{2}$ cups plain rice
2$\frac{1}{2}$ cups hot water
1 tsp. salt
2 bay leaves
1 cup dry vermouth

Heat oven to 300°. In a 4-quart ovenproof saucepan or casserole with a lid, melt butter and cook onions for 8 minutes. Add remaining ingredients and cover tightly. Bake for 15 to 20 minutes or until liquid is absorbed and rice is done. Remove bay leaves and fluff with a fork.

WILD AND WHITE RICE PILAF

You can add or delete ingredients to complement your main dish. Dried fruits enhance poultry or pork, while the vegetables go well with beef or fish. Cooking rice in the convection oven is easier as well as time-saving.

1/4 cup butter
2 onions, finely chopped
2 stalks celery, finely chopped
2 cloves garlic, minced
3 1/2 cups chicken stock
1 cup wild rice
salt and pepper
1 cup white rice

ADDITIONS #1

1 cup golden raisins
1/4 cup chopped dried apricots
1/2 cup chopped pitted prunes
3/4 cup finely chopped fresh parsley
1/2 cup slivered almonds, toasted

ADDITIONS #2

1 red bell pepper, finely diced
2 carrots, finely chopped
3/4 cup finely chopped fresh parsley
3/4 cup pine nuts, toasted

Heat oven to 300°. In a 4-quart ovenproof saucepan or casserole with a lid, melt butter. Add onion, celery, and garlic and cook for 10 minutes. Add chicken stock, wild rice, salt and pepper. Cover and cook for 20 minutes. Add white rice and addition of your choice, except for nuts and parsley, and cook for 15 to 20 minutes longer. Test for doneness. Stir in parsley and toasted nuts.

CONFETTI GRITS

Use any combination of leftover cheeses you have on hand. It's hard to go wrong with this recipe. This makes an excellent side dish for barbecues, buffets or brunches.

1 cup chopped ham
1 green bell pepper, cored and
  chopped
1 red bell pepper, cored and chopped
1 onion, chopped
1 clove garlic, minced
1/4 cup butter

4 cups water
1 cup grits
1 tsp. seasoning salt
2 eggs, beaten
2 cups shredded cheese: cheddar,
  Swiss, Jack, Havarti or any
  combination

Sauté ham, peppers, onion and garlic in butter. Set aside to cool. In a saucepan, bring water to a boil, stir in grits and cook over medium heat until thickened, about 5 minutes. Add ham mixture, seasoning salt, eggs and cheese. Pour into a greased casserole. May be covered and refrigerated for several days at this point. To bake, heat oven to 300° and bake for 30 to 40 minutes until bubbly.

BUFFET POTATO BAKE

These are ideal for a buffet, because they can be made ahead and go with almost anything. People love them, so be sure to make a lot.

1 pkg. (20 oz.) frozen shredded potatoes for hash browns
1 can (10. oz.) cream of chicken soup
1/4 cup butter
1 pt. sour cream
1 1/2 cups shredded cheddar cheese
1/2 cup sliced green onions
1/2 cup cornflake crumbs or crushed potato chips

Thaw potatoes. In a large bowl, combine soup, butter, sour cream, cheese and green onions. Stir in potatoes. Pour into a greased ovenproof casserole. Top with crumbs. Bake at 325° for 45 to 60 minutes or until bubbly and golden.

SWEET POTATOES WITH PEACHES AND CASHEWS

If you're looking for something different to do with sweet potatoes that isn't too sweet, this is a great recipe.

1 can (16 oz.) sweet potatoes, drained and sliced
1 can (16 oz.) sliced peaches, drained
1/3 cup brown sugar, packed
1/2 cup cashew pieces
1/2 tsp. salt
1/2 tsp. ground ginger
2 tbs. butter, melted

Heat oven to 325°. Place sweet potatoes in a buttered casserole dish and combine with sliced peaches. In a small bowl, combine brown sugar, cashews, salt and ginger. Sprinkle over sweet potatoes and drizzle with butter. Bake for 30 minutes, or until heated through and bubbly.

SCALLOPED SWEET POTATOES
WITH CRANBERRIES AND APPLES

Unusual, colorful, easy and delicious — what more could you ask? Ideal for holiday entertaining when you want a dish that travels well or is easy to multiply.

2 cans (18 oz. each) drained sweet potatoes or yams
1 can (21 oz.) apple pie filling
1 can (18 oz.) whole cranberries
2 tbs. orange marmalade
2 tbs. apricot jam

In a colorful ovenproof casserole, approximately 8 inches square, place sweet potatoes, apple pie filling and whole cranberries; gently stir together. Combine marmalade and apricot jam and spread over top. Bake at 325° for 20 to 30 minutes or until bubbly.

PASTA SOUFFLÉ

Serve this rich, delicious soufflé with plain roasted chicken, veal or beef.

4 cups water
1 chicken bouillon cube (prefer Knorr)
1 pkg. (8 oz.) very fine egg noodles
1/2 cup butter, melted
4 oz. cream cheese, room temperature
3 eggs, beaten
1 tbs. lemon juice
1 tsp. sugar
1 tsp. Worcestershire sauce
salt and pepper

Heat oven to 300°. Bring water to a boil in a large saucepan. Add bouillon cube and cook noodles until tender. Drain and rinse well with cold water. In a bowl, combine butter, cream cheese, eggs, lemon juice, sugar, Worcestershire and seasonings. Stir in noodles. Pour into a buttered 1 1/2-quart soufflé dish or a casserole. Bake for 45 minutes or until puffed and golden.

DESSERTS

IVORY CHEESECAKE

This is truly an impressive dessert. You can change the liquor and sauce for a variety of flavors. Try Captain Whidbey's Loganberry Liqueur and raspberry sauce; Grand Marnier and strawberry sauce; amaretto and apricot sauce; or Bailey's Irish Cream and chocolate sauce.

CRUST

1³/₄ cups graham cracker crumbs
6 tbs. butter, melted
2 tbs. sugar
2 tsp. vanilla extract

Butter a 9-x-3-inch springform pan. Combine crust ingredients and press into bottom and sides of pan.

FILLING

3 lb. cream cheese, room temperature
5 eggs
1 1/2 lb. white chocolate, melted
1/4 cup flour
1/4 cup sugar
1 1/2 cups sour cream
1/4 cup liqueur of your choice

Heat oven to 325°. Fill a 9-inch cake pan with very hot water and set on the bottom rack. Bake cheesecake on top rack.

With a mixer or food processor, combine cream cheese and eggs until smooth. Add remaining ingredients and mix until combined. Pour into prepared crust. Place in oven and bake for 30 minutes. Reduce oven heat to 300° and bake for an additional 30 minutes. If top begins to brown, cover with foil and secure tightly. Turn oven off and leave cake in oven for 25 additional minutes. Cool and refrigerate for several hours or overnight. Two hours before serving, remove cake from the refrigerator. Top or serve with desired sauce.

APPLE CAKE WITH WHISKEY GLAZE

Heavy, moist and loaded with apples, this cake keeps well — if you hide it!

3 cups finely diced peeled apples
2 cups chopped walnuts or pecans
1 1/2 cups raisins
2 cups sugar
1 1/2 cups vegetable oil

3 eggs
3 cups flour
1 tsp. baking soda
1 tsp. salt
1 tsp. cinnamon

Heat oven to 325°. Combine apples, nuts and raisins in a large bowl. In a mixer or food processor, combine sugar, oil and eggs until smooth. Add flour, soda, salt and cinnamon. Process to combine. Add to apple mixture and stir well. Grease and flour a Bundt or 10-inch tube pan. Spoon batter into pan. Bake for 45 minutes to 1 hour or until cake tests done with a toothpick. Remove from oven and cool slightly.

GLAZE

1 cup brown sugar, packed
1/2 cup butter

1/2 cup cream
1/4 cup whiskey or bourbon, or vanilla
    extract

Combine glaze ingredients in a small saucepan and bring to a boil. Poke holes in cake with a fork and pour over cake. Let stand for several hours before serving.

HUMMINGBIRD CAKE

Where this recipe got its name is anybody's guess. It's nice for potlucks and church functions.

2 cups sugar
3 eggs
1 1/2 cups vegetable oil
1 1/2 tsp. vanilla extract
2 cups mashed bananas
1 can (8 oz.) crushed pineapple,
   undrained

3 cups flour
1 tsp. baking soda
1 tsp. salt
2 tsp. cinnamon
2 cups chopped walnuts

Heat oven to 325°. With an electric mixer, combine sugar, eggs, oil and vanilla until smooth. Add bananas and crushed pineapple. Stir in dry ingredients and nuts. Pour into a well-greased Bundt or 10-inch tube pan. Bake for 20 to 30 minutes or until cake tests done. Remove from pan and cool. Frost with cream cheese icing and garnish with additional chopped walnuts.

LEMON CREAM CHEESE TART

Servings: 8

Using fresh lemon juice and lots of grated zest really makes the flavor sing.

CRUST

¾ cup butter

½ cup confectioners' sugar

1¼ cups flour

Heat oven to 300°. Combine ingredients with a mixer or food processor. Press dough into bottom and up sides of a 10-inch tart pan with a removable bottom. Prick bottom with a fork. Bake for 12 to 15 minutes or until a very light golden color. Cool.

FILLING

3 whole eggs

3 egg yolks

¾ cup sugar

1 tbs. arrowroot

¾ cup freshly squeezed lemon juice

1½ tsp. grated lemon zest

4 oz. cream cheese, room temperature

Heat oven to 300°. With a mixer or food processor, combine filling ingredients. Pour into prepared crust. Bake for 15 to 20 minutes or until filling is set. Cool.

HONEY ALMOND TART

If you have a sweet tooth, this one's for you! The crust is also suitable for other tarts as well—just bake and fill as desired.

CRUST

¾ cup butter   1¼ cups flour
½ cup confectioners' sugar

Heat oven to 300°. Combine ingredients with a mixer or food processor. Press dough into bottom and up sides of a 10-inch tart pan with a removable bottom. Prick bottom with a fork. Bake for 12 to 15 minutes or until very light golden. Cool.

FILLING

1 cup sugar   8 oz. sliced almonds
1 cup honey   2 tsp. almond extract
1 cup butter, melted

Combine all ingredients in a saucepan and melt together over medium heat, stirring constantly, until sugar is dissolved. Pour into cooled crust. Bake at 300° for 15 minutes or until set. Cool completely before cutting.

MACADAMIA NUT DESSERT

This is a perfect ending for a Luau Pork dinner (page 87). It's very rich, but we still like to gild the lily with a bit of whipped cream. Serve Kona coffee to keep in the spirit of things.

4 eggs
³/₄ cup sugar
1¹/₃ cups light corn syrup
3 tbs. dark rum, prefer Meyers
1 tsp. vanilla extract
1¹/₂ cups chopped macadamia nuts

Heat oven to 350°. With a mixer or food processor, combine all ingredients. Pour into a buttered 9-inch pie plate. Bake for 30 to 40 minutes or until set.

EASY SOUFFLÉ GRAND MARNIER

This is not actually a soufflé, but it sure acts like one. Don't tell anyone how easy it is and we guarantee they'll be impressed.

1 pkg. (8 oz.) cream cheese, cut into cubes
5 eggs
½ cup sugar
¾ cup whipping cream
¼ cup Grand Marnier liqueur
1 pkg. (10 oz.) frozen raspberries, thawed

Heat oven to 350°. With a mixer or food processor, combine cream cheese with eggs until mixture is smooth. Add sugar, cream and liqueur and blend until smooth. Butter a 2-quart soufflé, dish and sprinkle bottom and sides with sugar to coat. This will give the soufflé something to climb up on. Bake for 35 to 45 minutes or until light golden brown and puffed. Spoon into dessert dishes and spoon raspberries over top. Soufflé may be kept in refrigerator for several hours before baking. Remove 15 minutes prior to placing in oven.

PUMPKIN DESSERT

Here's a quick and easy dessert for a crowd around holiday time. Serve with whipped cream.

1 can (29 oz.) pumpkin
3 eggs, beaten
1 1/4 cups sugar
2 tsp. cinnamon
1/2 tsp. ground ginger
1/4 tsp. ground cloves
1/4 tsp. salt
3/4 cup evaporated milk
1 pkg. (18 oz.) yellow cake mix
1 cup butter, melted
1 cup chopped walnuts or pecans

Heat oven to 325°. With an electric mixer or food processor, combine pumpkin, eggs, sugar, spices and evaporated milk until smooth. Pour into a buttered ovenproof casserole. Sprinkle dry cake mix over top, drizzle with melted butter and sprinkle with nuts. Bake for 45 minutes or until golden and custard is set. Cool.

BANANAS JERICHO

Quick and glamorous, this is fun to serve for company. Bake in a pretty quiche dish and flame at the table.

$1/4$ cup honey
dash cinnamon
4 bananas, peeled and sliced into 1-inch pieces
$1/4$ cup crème de banana liqueur
$1/4$ cup brandy
1 pt. rich vanilla ice cream

Heat oven to 325°. Place honey in a dish, sprinkle with cinnamon and add sliced bananas. Drizzle with banana liqueur and turn to coat. Bake bananas for 10 to 15 minutes or until they begin to soften. Heat brandy in a saucepan or in the microwave. Scoop ice cream into individual dishes. Pour brandy over bananas and ignite. Spoon sauce over ice cream and serve.

COCONUT BREAD PUDDING

You can purchase cream of coconut in the liquor section of the grocery store.

1 lb. loaf firm white bread
½ cup shredded coconut
2½ cups canned cream of coconut
2 cups whole milk
½ cup sugar
1 tbs. vanilla extract
6 eggs

Heat oven to 325°. Trim crusts from bread and cut bread into 1 inch cubes. Discard crusts. Toss bread and shredded coconut together and place in a buttered 2-quart casserole. Combine cream of coconut, milk, sugar, vanilla and eggs using an electric mixer or food processor. Pour egg mixture over bread cubes, pressing cubes down with a spoon to make sure they are covered with the milk mixture. Let stand 15 minutes.

Bake until pudding is set and top is golden, about 40 minutes. Serve warm.

CHEATER'S CRÈME BRULÉE

Servings: 6

This recipe is much lower in fat and calories than the traditional cream version. The taste is close to the original and you can enjoy it with less damage to your diet.

4 large egg yolks
1/2 cup nonfat sweetened condensed
   milk (green label)
1 tbs. cornstarch

2 1/2 cups 1% low-fat milk
1 tbs. vanilla extract
6 tbs. light brown sugar, packed

Heat oven to 325°. Place six 6 oz. ramekins or custard cups in a shallow baking pan. Heat a kettle of water for a water bath (bain marie). Whisk egg yolks, condensed milk and cornstarch until smooth. Heat milk over low heat or in a microwave until steaming. Gradually stir hot milk into egg mixture. Add vanilla. Skim foam. Pour mixture into ramekins. Place pan on center rack of oven and pour boiling water into the pan until it comes halfway up the sides. Bake for 30 to 40 minutes or until edges are set but center still wiggles.

Cool on a wire rack. Cover and chill for 2 hours or up to 2 days. To serve, divide brown sugar evenly on top of custards. Spread gently with the bottom of a spoon to sides of ramekins. Preheat broiler. Place custards back in the baking pan and surround with ice cubes (to prevent custards from overcooking). Broil until sugar melts and caramelizes. Be sure to watch closely! Serve immediately or chilled.

CHOCOLATE SOUFFLÉ

This sinfully rich dessert is one of our favorites. The center remains liquid and gooey and the edges are crispy and crunchy. Serve it warm topped with vanilla ice cream for a heavenly treat.

6 tbs. butter
4 oz. semi-sweet chocolate
$1/4$ cup sugar
$1^3/_4$ tbs. cornstarch
2 eggs and 2 egg yolks

Melt butter and chocolate together in the microwave or in a saucepan over low heat. Combine sugar and cornstarch in a medium bowl. Add chocolate mixture. Whisk eggs and egg yolks together in a separate bowl and add to chocolate mixture. Cover and refrigerate overnight.

Heat oven to 375°. Generously butter a 4-cup baking dish. Pour chocolate mixture into dish. Bake on top oven rack for 15 to 20 minutes. Scoop into individual dessert bowls and serve immediately with ice cream.

COCONUT LEMON BARS

Really a different combination of flavors and textures — the lemon zest gives a nice tang. These freeze well.

CRUST

½ cup butter   1 cup flour
½ cup brown sugar, packed

Heat oven to 325°. Combine ingredients and press to cover the bottom of an ungreased 9-inch square pan. Bake for 10 minutes. Cool a few minutes before spreading with topping.

TOPPING

2 eggs, beaten   ½ tsp. salt
1 cup brown sugar, packed   1 cup coconut
1 tsp. lemon zest   1 cup chopped walnuts
2 tbs. lemon juice   ½ cup dark raisins

Combine ingredients and spread on top of crust. Bake for 15 to 20 minutes or until center of filling is set. Cool and cut into bars.

BUTTERSCOTCH BROWNIES

Chewy and delicious! If you wish, add chocolate chips to the recipe. Take care not to overbake, as they are much better underdone.

1/2 cup butter, room temperature
1 cup brown sugar, packed
1 egg
1 1/2 tsp. vanilla extract
1 cup flour
1/2 tsp. baking soda
1/2 tsp. baking powder
pinch salt
1 cup chopped pecans
1 cup chocolate chips, optional

Heat oven to 300°. Grease an 8-inch baking pan. Combine butter, sugar, egg and vanilla until light and fluffy. Add flour, baking soda, baking powder and salt. Stir in nuts and chips, if desired. Spread batter in prepared pan. Bake for 20 minutes or until beginning to pull away from sides of pan. Cool and cut into 2-inch squares.

INDEX

Serve Creative, Easy, Nutritious Meals with **nitty gritty®** Cookbooks

1 or 2, Cooking for
100 Dynamite Desserts
9 x 13 Pan Cookbook
Bagels, Best
Barbecue Cookbook
Beer and Good Food
Big Book of Bread Machine Recipes
Big Book of Kitchen Appliance Recipes
Big Book of Snacks and Appetizers
Blender Drinks
Bread Baking
Bread Machine
Bread Machine II
Bread Machine III
Bread Machine V
Bread Machine VI
Bread Machine, Entrees
Burger Bible
Cappuccino/Espresso
Casseroles
Chicken, Unbeatable
Chile Peppers
Clay, Cooking in

Coffee and Tea
Convection Oven
Cook-Ahead Cookbook
Crockery Pot, Extra-Special
Deep Fryer
Dehydrator Cookbook
Edible Gifts
Edible Pockets
Fabulous Fiber Cookery
Fondue and Hot Dips
Fondue, New International
Freezer, 'Fridge, Pantry
Garlic Cookbook
Grains, Cooking with
Healthy Cooking on Run
Ice Cream Maker
Indoor Grill, Cooking on
Italian Recipes, Quick and Easy
Juicer Book II
Kids, Cooking with Your
Kids, Healthy Snacks for
Loaf Pan, Recipes for
Low-Carb Recipes

Lowfat American
No Salt No Sugar No Fat   (REVISED)
Party Foods/Appetizers
Pasta Machine Cookbook
Pasta, Quick and Easy
Pinch of Time
Pizza, Best
Porcelain, Cooking in
Pressure Cooker, Recipes (REVISED)
Rice Cooker
Rotisserie Oven Cooking
Sandwich Maker
Simple Substitutions
Skillet, Sensational
Slow Cooking
Slow Cooker, Vegetarian
Soups and Stews
Soy & Tofu Recipes
Tapas Fantásticas
Toaster Oven Cookbook
Waffles & Pizzelles
Wedding Catering Cookbook
Wraps and Roll-Ups   (REVISED)

For a free catalog, call: Bristol Publishing Enterprises
(800) 346-4889
www.bristolpublishing.com